FREE DVD FREE FREE DVD

Essential Test Tips DVD from Trivium Test Prep

Dear Customer,

Thank you for purchasing from Cirrus Test Prep! Whether you're looking to join the military, get into college, or advance your career, we're honored to be a part of your journey.

To show our appreciation (and to help you relieve a little of that test-prep stress), we're offering a **FREE *TExES Essential Test Tips DVD*** by Cirrus Test Prep. Our DVD includes 35 test preparation strategies that will help keep you calm and collected before and during your big exam. All we ask is that you email us your feedback and describe your experience with our product. Amazing, awful, or just so-so: we want to hear what you have to say!

To receive your **FREE *TExES Essential Test Tips DVD***, please email us at 5star@cirrustestprep.com. Include "Free 5 Star" in the subject line and the following information in your email:

1. The title of the product you purchased.

2. Your rating from 1 – 5 (with 5 being the best).

3. Your feedback about the product, including how our materials helped you meet your goals and ways in which we can improve our products.

4. Your full name and shipping address so we can send your **FREE *TExES Essential Test Tips DVD***.

If you have any questions or concerns please feel free to contact us directly at 5star@cirrustestprep.com.

Thank you, and good luck with your studies!

* Please note that the free DVD is <u>not included</u> with this book. To receive the free DVD, please follow the instructions above.

TExES Mathematics 7–12 (235) Study Guide:

TEST PREP WITH 400+ MATH PRACTICE QUESTIONS FOR THE TEXAS EXAMINATIONS OF EDUCATOR STANDARDS (235) [6TH EDITION]

J.G. COX

Table of Contents

Online Resources

To help you fully prepare for your TExES Mathematics 7 – 12 (235) test, Cirrus includes online resources with the purchase of this study guide.

PRACTICE TEST

In addition to the practice test included in this book, we also offer an online exam. Since many exams today are computer-based, getting to practice your test-taking skills on the computer is a great way to prepare.

FLASHCARDS

A convenient supplement to this study guide, Cirrus's e-flashcards enable you to review important terms easily on your computer or smartphone.

CHEAT SHEETS

Review the skills you need to master with easy-to-read Cheat Sheets. Topics covered include Numbers and Operations, Algebra, Geometry, Statistics, and Probability.

FROM STRESS TO SUCCESS

Watch *From Stress to Success*, a brief but insightful YouTube video that offers the tips, tricks, and secrets experts use to score higher on the exam.

REVIEWS

Leave a review, send us helpful feedback, or sign up for Cirrus's promotions—including free books!

To access these materials, please enter the following URL into your browser: **www.cirrustestprep.com/texes-math-online-resources**.

Introduction

Congratulations on choosing to take the TExES Mathematics 7 – 12 (235) test! By purchasing this book, you've taken the first step toward becoming a math teacher.

This guide will provide you with a detailed overview of the TExES Mathematics 7 – 12 test, so you know exactly what to expect on test day. We'll take you through all the concepts covered on the test and give you the opportunity to test your knowledge with practice questions. Even if it's been a while since you last took a major test, don't worry; we'll make sure you're more than ready!

WHAT IS THE TEXES MATHEMATICS TEST?

The TExES Mathematics 7 – 12 test measures aptitude in mathematics for teacher candidates looking to certify as math teachers. This test must be taken *in addition to* the assessments in reading, writing, mathematics, and professional knowledge required in Texas. The TExES Mathematics 7 – 12 test does not replace these other exams.

WHAT'S ON THE TEXES MATHEMATICS TEST?

The TExES Mathematics 7 – 12 test gauges college-level content knowledge in mathematics, as well as the necessary skills for mathematics. Candidates are expected to demonstrate thorough and extensive conceptual knowledge of subjects including number sense, algebra, geometry, trigonometry, and calculus. You will also be expected to demonstrate mastery of key skills related to probability and statistics, mathematical reasoning, and best practices in mathematics instruction and assessment. The content is organized into six domains, which are further divided into twenty-one competencies.

You will have five hours to answer 100 multiple-choice questions.

What's on the TExES Mathematics 7 – 12 (235) Test?

Domains	Competencies	Percentage
I. Number Concepts	1. Real number system 2. Complex number system 3. Number theory concepts and principles	14%
II. Patterns and Algebra	4. Use of patterns to model and solve problems 5. Functions, relations, and their graphs 6. Linear and quadratic functions 7. Polynomial, rational, radical, absolute value, and piecewise functions 8. Exponential and logarithmic functions 9. Trigonometric and circular functions 10. Differential and integral calculus	33%
III. Geometry and Measurement	11. Understanding measurement as a process 12. Axiomatic nature of geometries (especially Euclidean geometry) 13. Results, uses, and applications of Euclidean geometry 14. Coordinate, transformational, and vector geometry	19%
IV. Probability and Statistics	15. Graphical and numerical techniques for exploring data and recognizing patterns 16. Concepts and applications of probability 17. Relationships among probability theory, sampling, and statistical interference; using statistical interference to make and evaluate predictions	14%
V. Mathematical Processes and Perspectives	18. Mathematical reasoning and problem solving 19. Intra- and interdisciplinary mathematical connections; communicating mathematical concepts	10%
VI. Mathematics Learning, Instruction, and Assessment	20. Planning, organization, and implementation of instruction to best match how children learn mathematics in accordance with statewide curriculum 21. Assessment	10%

Domain 1 assesses your understanding of numbers: the different numbering systems and the fundamentals of number theory. You must be able to use basic operations on different types of numbers and to prove results. You also must be able to apply your understanding of number systems and theory to solve problems.

Domain II addresses your understanding of the fundamentals of algebra, including symbolic reasoning and fundamental algebraic structures. You must be able to demonstrate mastery of linear and nonlinear equations and inequalities, rational and radical expressions, and polynomial equations. It also addresses the structures and concepts underlying trigonometry and calculus. You should be able to prove and apply the Pythagorean theorem, as well as all of the fundamental formulas of trigonometry. In addition, you should also be able to demonstrate understanding of the rules of differentiation, the proper ways to use derivatives, and be able to derive, prove, and apply the fundamental concepts of integrals, sequences, and series. Finally, you must be able to apply algebraic and calculus skills to solve problems.

Domain III focuses on the study of the properties and relationships between objects in space and the study of mathematical structures. You must be able to describe, measure, and analyze these objects and structures using appropriate mathematical processes and tools within the framework of Euclidean geometry. In particular, you must demonstrate a mastery of the theorems that prove the geometric principles. Finally, you must demonstrate understanding of alternate geometric systems: transformational, coordinate, and vector.

Domain IV measures your ability to effectively collect, manage, analyze, interpret, and represent data. It also assesses your understanding of probability, including basic principles, and the differences among finite probability, conditional probability, and independence. Finally, it assesses your understanding of the relationship among probability, sampling, and statistical interference. You also should be able to apply your skills in statistics and probability to problem solving.

Domains V and VI move away from specific mathematical content and instead address mathematical and instructional skills. You must be able to demonstrate a mastery of logic: both its major concepts and its application. You also must be able to make instructional choices that best promote student learning by properly sequencing lessons and identifying and addressing misconceptions through both classroom activities and assessments.

How is the TExES Mathematics Test Scored?

Your scores on your TExES Mathematics 7 – 12 test will become available online seven days after your test date. For more information, check cms.texes-ets.org. Your scores will be automatically available to the Texas Education Agency (TEA). However, to have them sent to a college or university, you must make a request when you register for the exam.

Each multiple-choice question is worth one raw point. The total number of questions you answer correctly is added up to obtain your raw score, which is then converted to a scale of 100–300. You must receive a score of 240 to pass the TExES Mathematics 7 – 12 test.

There will be some questions on the test that are not scored; however, you will not know which ones these are. ETS uses these to test out new questions for future exams.

There is no guess penalty on the TExES, so you should always guess if you do not know the answer to a question.

HOW IS THE TExES MATHEMATICS TEST ADMINISTERED?

The TExES Mathematics 7 – 12 test is a computer-based test offered continuously by appointment at a range of universities and testing centers. Check cms.texes-ets. org for more information. You will need to print your registration ticket from your online account and bring it, along with your identification, to the testing site on test day. No pens, pencils, erasers, printed or written materials, or electronic devices are allowed. **An on-screen graphing calculator will be provided. You may not bring a calculator into the testing site.** You also may not bring any kind of bag into the testing room or wear headwear (unless for religious purposes). You may take the test once every forty-five days.

ABOUT CIRRUS TEST PREP

Cirrus Test Prep study guides are designed by current and former educators and are tailored to meet your needs as an incoming educator. Our guides offer all of the resources necessary to help you pass teacher certification tests across the nation.

Cirrus clouds are graceful, wispy clouds characterized by their high altitude. Just like cirrus clouds, Cirrus Test Prep's goal is to help educators "aim high" when it comes to obtaining their teacher certification and entering the classroom.

ABOUT THIS GUIDE

This guide will help you master the most important test topics and also develop critical test-taking skills. We have built features into our books to prepare you for your tests and increase your score. Along with a detailed summary of the test's format, content, and scoring, we offer an in-depth overview of the content knowledge required to pass the test. Our sidebars provide interesting information, highlight key concepts, and review content so that you can solidify your understanding of the exam's concepts. Test your knowledge with sample questions and detailed answer explanations in the text that help you think through the problems on the exam and practice questions that reflect the content and format of the TExES Mathematics 7 – 12 test. We're pleased you've chosen Cirrus to be a part of your professional journey!

Numbers and Operations

This chapter provides a review of the basic yet critical components of mathematics such as manipulating fractions, comparing numbers, and using units. These concepts will provide the foundation for more complex mathematical operations in later chapters.

TYPES OF NUMBERS

Numbers are placed in categories based on their properties.

- A **natural number** is greater than 0 and has no decimal or fraction attached. These are also sometimes called counting numbers {1, 2, 3, 4, ...}.

- **Whole numbers** are natural numbers and the number 0 {0, 1, 2, 3, 4, ...}.

- **Integers** include positive and negative natural numbers and 0 {..., −4, −3, −2, −1, 0, 1, 2, 3, 4, ...}.

- A **rational number** can be represented as a fraction. Any decimal part must terminate or resolve into a repeating pattern. Examples include −12, $-\frac{4}{5}$, 0.36, 7.$\overline{7}$, 26$\frac{1}{2}$, etc.

- An **irrational number** cannot be represented as a fraction. An irrational decimal number never ends and never resolves into a repeating pattern. Examples include $-\sqrt{7}$, π, and 0.34567989135...

- A **real number** is a number that can be represented by a point on a number line. Real numbers include all the rational and irrational numbers.

- An **imaginary number** includes the imaginary unit i, where $i = \sqrt{-1}$. Because $i^2 = -1$, imaginary numbers produce a negative value when squared. Examples of imaginary numbers include $-4i$, $0.75i$, $i\sqrt{2}$ and $\frac{8}{3}i$.

▸ A **complex number** is in the form $a + bi$, where a and b are real numbers. Examples of complex numbers include $3 + 2i$, $-4 + i$, $\sqrt{3} - i\sqrt[3]{5}$ and $\frac{5}{8} - \frac{7i}{8}$. All imaginary numbers are also complex.

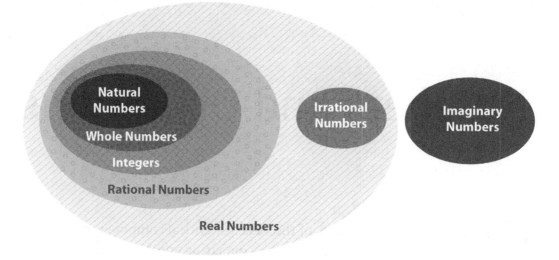

Figure 1.1. Types of Numbers

The **factors** of a natural number are all the numbers that can multiply together to make the number. For example, the factors of 24 are 1, 2, 3, 4, 6, 8, 12, and 24. Every natural number is either prime or composite. A **prime number** is a number that is only divisible by itself and 1. (The number 1 is not considered prime.) Examples of prime numbers are 2, 3, 7, and 29. The number 2 is the only even prime number. A **composite number** has more than two factors. For example, 6 is composite because its factors are 1, 6, 2, and 3. Every composite number can be written as a unique product of prime numbers, called the **prime factorization** of the number. For example, the prime factorization of 90 is $90 = 2 \times 3^2 \times 5$. All integers are either even or odd. An even number is divisible by 2; an odd number is not.

HELPFUL HINT

If a real number is a natural number (e.g., 50), then it is also a whole number, an integer, and a rational number.

PROPERTIES OF NUMBER SYSTEMS

A system is **closed** under an operation if performing that operation on two elements of the system results in another element of that system. For example, the integers are closed under the operations of addition, subtraction, and multiplication but not division. Adding, subtracting, or multiplying two integers results in another integer. However, dividing two integers could result in a rational number that is not an integer $\left(-2 \div 3 = \frac{-2}{3}\right)$.

- ▶ The rational numbers are closed under all four operations (except for division by 0).
- ▶ The real numbers are closed under all four operations.
- ▶ The complex numbers are closed under all four operations.
- ▶ The irrational numbers are NOT closed under ANY of the four operations.

The **commutative property** holds for an operation if order does not matter when performing the operation. For example, multiplication is commutative for integers: $(-2)(3) = (3)(-2)$.

The **associative property** holds for an operation if elements can be regrouped without changing the result. For example, addition is associative for real numbers: $-3 + (-5 + 4) = (-3 + -5) + 4$.

The **distributive property** of multiplication over addition allows a product of sums to be written as a sum of products: $a(b + c) = ab + ac$. The value a is distributed over the sum $(b + c)$. The acronym FOIL (First, Outer, Inner, Last) is a useful way to remember the distributive property.

When an operation is performed with an **identity element** and another element a, the result is a. The identity element for multiplication on real numbers is 1 ($a \times 1 = a$), and for addition is 0 ($a + 0 = a$).

An operation of a system has an **inverse element** if applying that operation with the inverse element results in the identity element. For example, the inverse element of a for addition is $-a$ because $a + (-a) = 0$. The inverse element of a for multiplication is $\frac{1}{a}$ because $a \times \frac{1}{a} = 1$.

SAMPLE QUESTIONS

1) Classify the following numbers as natural, whole, integer, rational, or irrational. (The numbers may have more than one classification.)

   A.   72

   B. $-\frac{2}{3}$

   C. $\sqrt{5}$

   Answers:

   A.   The number is **natural**, **whole**, an **integer**, and **rational**.

   B.   The fraction is **rational**.

   C.   The number is **irrational**. (It cannot be written as a fraction, and written as a decimal is approximately 2.2360679...)

2) Determine the real and imaginary parts of the following complex numbers.

   A.   20

   B. $10 - i$

   C. $15i$

   Answers:

   A complex number is in the form of $a + bi$, where a is the real part and bi is the imaginary part.

   A. $20 = 20 + 0i$
         The real part is 20, and there is no imaginary part.

   B. $10 - i = 10 - 1i$
         The real part is 10, and $-1i$ is the imaginary part.

   C. $15i = 0 + 15i$
         The real part is 0, and the imaginary part is 15i.

3) Answer True or False for each statement:

   A.   The natural numbers are closed under subtraction.

   B.   The sum of two irrational numbers is irrational.

   C.   The sum of a rational number and an irrational number is irrational.

   Answers:

   A.   **False**. Subtracting the natural number 7 from 2 results in $2 - 7 = -5$, which is an integer, but not a natural number.

   B.   **False**. For example, $(5 - 2\sqrt{3}) + (2 + 2\sqrt{3}) = 7$. The sum of two irrational numbers in this example is a whole number, which is not irrational. The sum of a rational number and an irrational number is sometimes rational and sometimes irrational.

   C.   **True**. Because irrational numbers have decimal parts that are unending and with no pattern, adding a repeating or terminating decimal will still result in an unending decimal without a pattern.

4) Answer True or False for each statement:

   A.   The associative property applies for multiplication in the real numbers.

   B.   The commutative property applies to all real numbers and all operations.

   Answers:

   A.   **True**. For all real numbers, $a \times (b \times c) = (a \times b) \times c$. Order of multiplication does not change the result.

   B.   **False**. The commutative property does not work for subtraction or division on real numbers. For example, $12 - 5 = 7$, but $5 - 12 = -7$ and $10 \div 2 = 5$, but $2 \div 10 = \frac{1}{5}$.

OPERATIONS WITH COMPLEX NUMBERS

Operations with complex numbers are similar to operations with real numbers in that complex numbers can be added, subtracted, multiplied, and divided. When adding or subtracting, the imaginary parts and real parts are combined separately. When multiplying, the distributive property (FOIL) can be applied. Note that multiplying complex numbers often creates the value i^2 which can be simplified to –1.

To divide complex numbers, multiply both the top and bottom of the fraction by the **complex conjugate** of the divisor (bottom number). The complex conjugate is the complex number with the sign of the imaginary part changed. For example, the complex conjugate of $3 + 4i$ would be $3 - 4i$. Since both the top and the bottom of the fraction are multiplied by the same number, the fraction is really just being multiplied by 1. When simplified, the denominator of the fraction will now be a real number.

SAMPLE QUESTIONS

5)   Simplify: $(3 - 2i) - (-2 + 8i)$

Answer:

$(3 - 2i) - (-2 + 8i)$	
$= (3 - 2i) - 1(-2 + 8i)$	Distribute the –1.
$= 3 - 2i + 2 - 8i$	
$= \mathbf{5 - 10i}$	Combine like terms.

6)   Simplify: $\dfrac{4i}{(5 - 2i)}$

Answer:

$\dfrac{4i}{(5 - 2i)}$	
$= \dfrac{4i}{5 - 2i}\left(\dfrac{5 + 2i}{5 + 2i}\right)$	Multiply the top and bottom of the fraction by the complex conjugate of $5 + 2i$.
$= \dfrac{20i + 8i^2}{25 + 10i - 10i - 4i^2}$	
$= \dfrac{20i + 8(-1)}{25 + 10i - 10i - 4(-1)}$	Simplify the result using the identity $i^2 = -1$.
$= \dfrac{20i - 8}{25 + 10i - 10i + 4}$	
$= \dfrac{20i - 8}{29}$	Combine like terms.
$= \dfrac{-8}{29} + \dfrac{20}{29}i$	Write the answer in the form $a + bi$.

SCIENTIFIC NOTATION

Scientific notation is a method of representing very large and small numbers in the form $a \times 10^n$, where a is a value between 1 and 10, and n is a nonzero integer. For example, the number 927,000,000 is written in scientific notation as 9.27×10^8. Multiplying 9.27 by 10 eight times gives 927,000,000. When performing operations with scientific notation, the final answer should be in the form $a \times 10^n$.

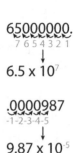

Figure 1.2. Scientific Notation

When adding and subtracting numbers in scientific notation, the power of 10 must be the same for all numbers. This results in like terms in which the a terms are added or subtracted and the 10^n remains unchanged. When multiplying numbers in scientific notation, multiply the a factors, and then multiply that answer by 10 to the sum of the exponents. For division, divide the a factors and subtract the exponents.

HELPFUL HINT

When multiplying numbers in scientific notation, add the exponents. When dividing, subtract the exponents.

SAMPLE QUESTIONS

7)   **Simplify: $(3.8 \times 10^3) + (4.7 \times 10^2)$**

Answer:

$(3.8 \times 10^3) + (4.7 \times 10^2)$	
$3.8 \times 10^3 = 3.8 \times 10 \times 10^2 = 38 \times 10^2$	To add, the exponents of 10 must be the same.
$38 \times 10^2 + 4.7 \times 10^2 = 42.7 \times 10^2$	Add the a terms together.
$= 4.27 \times 10^3$	Write the number in proper scientific notation.

8)   **Simplify: $(8.1 \times 10^{-5})(1.4 \times 10^7)$**

Answer:

$(8.1 \times 10^{-5})(1.4 \times 10^7)$	
$8.1 \times 1.4 = 11.34$ $-5 + 7 = 2$ $= 11.34 \times 10^2$	Multiply the a factors and add the exponents on the base of 10.
$= 1.134 \times 10^3$	Write the number in proper scientific notation.

POSITIVE AND NEGATIVE NUMBERS

Positive numbers are greater than 0, and **negative numbers** are less than 0. Both positive and negative numbers can be shown on a **number line**.

Figure 1.3. Number Line

The **absolute value** of a number is the distance the number is from 0. Since distance is always positive, the absolute value of a number is always positive. The absolute value of *a* is denoted |*a*|. For example, |–2| = 2 since –2 is two units away from 0.

Positive and negative numbers can be added, subtracted, multiplied, and divided. The sign of the resulting number is governed by a specific set of rules shown in the table below.

Table 1.1. Operations with Positive and Negative Numbers

Adding Real Numbers		Subtracting Real Numbers	
Positive + Positive = Positive	7 + 8 = 15	Negative – Positive = Negative	–7 – 8 = –7 + (–8) = –15
Negative + Negative = Negative	–7 + (–8) = –15	Positive – Negative = Positive	7 – (–8) = 7 + 8 = 15
Negative + Positive OR Positive + Negative = Keep the sign of the number with larger absolute value	–7 + 8 = 1 7 + –8 = –1	Negative – Negative = Change the subtraction to addition and change the sign of the second number; then use addition rules.	–7 – (–8) = –7 + 8 = 1 –8 – (–7) = –8 + 7 = –1
Multiplying Real Numbers		**Dividing Real Numbers**	
Positive × Positive = Positive	8 × 4 = 32	Positive ÷ Positive = Positive	8 ÷ 4 = 2
Negative × Negative = Positive	–8 × (–4) = 32	Negative ÷ Negative = Positive	–8 ÷ (–4) = 2
Positive × Negative OR Negative × Positive = Negative	8 × (–4) = –32 –8 × 4 = –32	Positive ÷ Negative OR Negative ÷ Positive = Negative	8 ÷ (–4) = –2 –8 ÷ 4 = –2

SAMPLE QUESTIONS

9) **Add or subtract the following real numbers:**

    A. $-18 + 12$

    B. $-3.64 + (-2.18)$

    C. $9.37 - 4.25$

    D. $86 - (-20)$

Answers:

    A.   Since $|-18| > |12|$, the answer is negative: $|-18| - |12| = 6$. So the answer is **–6**.

    B.   Adding two negative numbers results in a negative number. Add the values: **–5.82**.

    C.   The first number is larger than the second, so the final answer is positive: **5.12**.

    D.   Change the subtraction to addition, change the sign of the second number, and then add: $86 - (-20) = 86 + (+20) = $ **106**.

10) **Multiply or divide the following real numbers:**

    A. $\left(\frac{10}{3}\right)\left(-\frac{9}{5}\right)$

    B. $\frac{-64}{-10}$

    C. $(2.2)(3.3)$

    D. $-52 \div 13$

Answers:

    A.   Multiply the numerators, multiply the denominators, and simplify: $\frac{-90}{15} = $ **–6**.

    B.   A negative divided by a negative is a positive number: **6.4**.

    C.   Multiplying positive numbers gives a positive answer: **7.26**.

    D.   Dividing a negative by a positive number gives a negative answer: **–4**.

ORDER OF OPERATIONS

The **order of operations** is simply the order in which operations are performed. **PEMDAS** is a common way to remember the order of operations:

1.	Parentheses		4.	Division
2.	Exponents		5.	Addition
3.	Multiplication		6.	Subtraction

Multiplication and division, and addition and subtraction, are performed together from left to right. So, performing multiple operations on a set of numbers is a four-step process:

1. P: Calculate expressions inside parentheses, brackets, braces, etc.
2. E: Calculate exponents and square roots.
3. MD: Calculate any remaining multiplication and division in order from left to right.
4. AS: Calculate any remaining addition and subtraction in order from left to right.

Always work from left to right within each step when simplifying expressions.

SAMPLE QUESTIONS

11) Simplify: $2(21 - 14) + 6 \div (-2) \times 3 - 10$

Answer:

$2(21 - 14) + 6 \div (-2) \times 3 - 10$	
$= 2(7) + 6 \div (-2) \times 3 - 10$	Calculate expressions inside parentheses.
$= 14 + 6 \div (-2) \times 3 - 10$ $= 14 + (-3) \times 3 - 10$ $= 14 + (-9) - 10$	There are no exponents or radicals, so perform multiplication and division from left to right.
$= 5 - 10$ $= \mathbf{-5}$	Perform addition and subtraction from left to right.

12) Simplify: $-(3)^2 + 4(5) + (5 - 6)^2 - 8$

Answer:

$-(3)^2 + 4(5) + (5 - 6)^2 - 8$	
$= -(3)^2 + 4(5) + (-1)^2 - 8$	Calculate expressions inside parentheses.
$= -9 + 4(5) + 1 - 8$	Simplify exponents and radicals.
$= -9 + 20 + 1 - 8$	Perform multiplication and division from left to right.
$= 11 + 1 - 8$ $= 12 - 8$ $= \mathbf{4}$	Perform addition and subtraction from left to right.

13) Simplify: $\dfrac{(7-9)^3 + 8(10-12)}{4^2 - 5^2}$

Answer:

$\dfrac{(7-9)^3 + 8(10-12)}{4^2 - 5^2}$	
$= \dfrac{(-2)^3 + 8(-2)}{4^2 - 5^2}$	Calculate expressions inside parentheses.
$= \dfrac{-8 + (-16)}{16 - 25}$	Simplify exponents and radicals.
$= \dfrac{-24}{-9}$	Perform addition and subtraction from left to right.
$= \dfrac{8}{3}$	Simplify.

UNITS OF MEASUREMENT

The standard units for the metric and American systems are shown below, along with the prefixes used to express metric units.

Table 1.2. Units and Conversion Factors

Dimension	American	SI
length	inch/foot/yard/mile	meter
mass	ounce/pound/ton	gram
volume	cup/pint/quart/gallon	liter
force	pound-force	newton
pressure	pound-force per square inch	pascal
work and energy	cal/British thermal unit	joule
temperature	Fahrenheit	kelvin
charge	faraday	coulomb

Table 1.3. Metric Prefixes

Prefix	Symbol	Multiplication Factor
tera	T	1,000,000,000,000
giga	G	1,000,000,000
mega	M	1,000,000
kilo	k	1,000
hecto	h	100
deca	da	10

Prefix	Symbol	Multiplication Factor
base unit	--	--
deci	d	0.1
centi	c	0.01
milli	m	0.001
micro	μ	0.000001
nano	n	0.000000001
pico	p	0.000000000001

Units can be converted within a single system or between systems. When converting from one unit to another unit, a conversion factor (a numeric multiplier used to convert a value with a unit to another unit) is used. The process of converting between units using a conversion factor is sometimes known as dimensional analysis.

Table 1.4. Conversion Factors

1 in. = 2.54 cm	1 lb. = 0.454 kg
1 yd. = 0.914 m	1 cal = 4.19 J
1 mi. = 1.61 km	$1\,^{\circ}F = \frac{9}{5}\,^{\circ}C + 32\,^{\circ}C$
1 gal. = 3.785 L	$1\,cm^3 = 1\,mL$
1 oz. = 28.35 g	1 hr = 3600 s

SAMPLE QUESTIONS

14) **Convert the following measurements in the metric system.**

   A. 4.25 kilometers to meters

   B. $8\,m^2$ to mm^2

   Answers:

   A. $4.25\ km \left(\frac{1000\ m}{1\ km} \right) = \textbf{4250 m}$

   B. $\frac{8\ m^2}{1} \times \frac{1000\ mm}{1\ m} \times \frac{1000\ mm}{1\ m} = \textbf{8,000,000 mm}^2$

   Since the units are square units (m^2), multiply by the conversion factor twice, so that both meters cancel.

15) **Convert the following measurements in the American system.**

   A.   12 feet to inches

   B.   7 yd² to ft²

   Answers:

   A. $12 \, \text{ft} \left(\frac{12 \, \text{in}}{1 \, \text{ft}} \right) = \textbf{144 in}$

   B. $7 \, \text{yd}^2 (3\text{ft}^2/1\text{yd}^2)(3\text{ft}^2/1\text{yd}^2) = \textbf{63 ft}^2$

   Since the units are square units (ft²), multiply by the conversion factor twice.

16) **Convert the following measurements in the metric system to the American system.**

   A.   23 meters to feet

   B.   10 m² to yd²

   Answers:

   A. $23 \, \text{m} \left(\frac{3.28 \, \text{ft}}{1 \, \text{m}} \right) = \textbf{75.44 ft}$

   B. $\frac{10 \, \text{m}^2}{1} \times \frac{1.094 \, \text{yd}}{1 \, \text{m}} \times \frac{1.094 \, \text{yd}}{1 \, \text{m}} = \textbf{11.97 yd}^2$

17) **Convert the following measurements in the American system to the metric system.**

   A.   8 in³ to milliliters

   B.   16 kilograms to pounds

   Answers:

   A. $8 \, \text{in}^3 \left(\frac{16.39 \, \text{ml}}{1 \, \text{in}^3} \right) = \textbf{131.12 mL}$

   B. $16 \, \text{kg} \left(\frac{2.2 \, \text{lb}}{1 \, \text{kg}} \right) = \textbf{35.2 lb}$

DECIMALS AND FRACTIONS

DECIMALS

A **decimal** is a number that contains a decimal point. A decimal number is an alternative way of writing a fraction. The place value for a decimal includes **tenths** (one place after the decimal), **hundredths** (two places after the decimal), **thousandths** (three places after the decimal), etc.

Table 1.5. Place Values		
1,000,000	10^6	millions
100,000	10^5	hundred thousands
10,000	10^4	ten thousands
1,000	10^3	thousands
100	10^2	hundreds
10	10^1	tens
1	10^0	ones
.		decimal
$\frac{1}{10}$	10^{-1}	tenths
$\frac{1}{100}$	10^{-2}	hundredths
1/1000	10^{-3}	thousandths

Decimals can be added, subtracted, multiplied, and divided:

▶ To add or subtract decimals, line up the decimal point and perform the operation, keeping the decimal point in the same place in the answer.

> **HELPFUL HINT**
>
> To determine which way to move the decimal after multiplying, remember that changing the decimal should always make the final answer smaller.

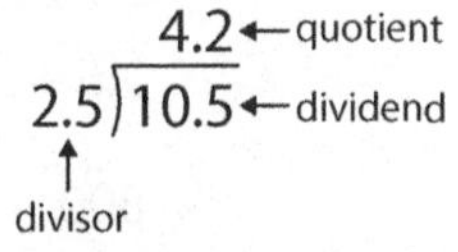

▶ To multiply decimals, first multiply the numbers without the decimal points. Then, sum the number of decimal places to the right of the decimal point in the original numbers and place the decimal point in the answer so that there are that many places to the right of the decimal.

▶ When dividing decimals move the decimal point to the right in order to make the divisor a whole number and move the decimal the same number of places in the dividend. Divide the numbers without regard to the decimal. Then, place the decimal point of the quotient directly above the decimal point of the dividend.

4.2 ←quotient
2.5)10.5 ←dividend
↑
divisor

Figure 1.4. Division Terms

SAMPLE QUESTIONS

18) Simplify: 24.38 + 16.51 − 29.87

Answer:

24.38 + 16.51 − 29.87	
24.38 + 16.51 = 40.89	Align the decimals and apply the order of operations left to right.
40.89 − 29.87 = **11.02**	

19) **Simplify: (10.4)(18.2)**

Answer:

(10.4)(18.2)	
104 × 182 = 18,928	Multiply the numbers ignoring the decimals.
18,928 → 189.28	The original problem includes two decimal places (one in each number), so move the decimal point in the answer so that there are two places after the decimal point.

Estimating is a good way to check the answer: $10.4 \approx 10$, $18.2 \approx 18$, and $10 \times 18 = 180$.

20) **Simplify: 80 ÷ 2.5**

Answer:

80 ÷ 2.5	
80 → 800 2.5 → 25	Move both decimals one place to the right (multiply by 10) so that the divisor is a whole number.
800 ÷ 25 = 32	Divide normally.

FRACTIONS

A **fraction** is a number that can be written in the form $\frac{a}{b}$, where b is not equal to 0. The a part of the fraction is the **numerator** (top number) and the b part of the fraction is the **denominator** (bottom number).

If the denominator of a fraction is greater than the numerator, the value of the fraction is less than 1 and it is called a **proper fraction** (for example, $\frac{3}{5}$ is a proper fraction). In an **improper fraction**, the denominator is less than the numerator and the value of the fraction is greater than 1 ($\frac{8}{3}$ is an improper fraction). An improper fraction can be written as a **mixed number**, which has a whole number part and a proper fraction part. Improper fractions can be converted to mixed numbers by dividing the numerator by the denominator, which gives the whole number part, and the remainder becomes the numerator of the proper fraction part. (For example, the improper fraction $\frac{25}{9}$ is equal to mixed number $2\frac{7}{9}$ because 9 divides into 25 two times, with a remainder of 7.)

Conversely, mixed numbers can be converted to improper fractions. To do so, determine the numerator of the improper fraction by multiplying the denominator by the whole number, and then adding the numerator. The final number is written as the (now larger) numerator over the original denominator.

Fractions with the same denominator can be added or subtracted by simply adding or subtracting the numerators; the denominator will remain unchanged. To add or subtract fractions with different denominators, find the **least common denominator (LCD)** of all the fractions. The LCD is the smallest number exactly divisible by each denominator. (For example, the least common denominator of the numbers 2, 3, and 8 is 24.) Once the LCD has been found, each fraction should be written in an equivalent form with the LCD as the denominator.

HELPFUL HINT

To convert mixed numbers to improper fractions:
$$a\frac{m}{n} = \frac{n \times a + m}{n}$$

To multiply fractions, the numerators are multiplied together and denominators are multiplied together. If there are any mixed numbers, they should first be changed to improper fractions. Then, the numerators are multiplied together and the denominators are multiplied together. The fraction can then be reduced if necessary. To divide fractions, multiply the first fraction by the reciprocal of the second.

HELPFUL HINT

$$\frac{a}{b} \pm \frac{c}{b} = \frac{a \pm c}{b}$$
$$\frac{a}{b} \times \frac{c}{d} = \frac{ac}{bd}$$
$$\frac{a}{b} \div \frac{c}{d} = \left(\frac{a}{b}\right)\left(\frac{d}{c}\right) = \frac{ad}{bc}$$

Any common denominator can be used to add or subtract fractions. The quickest way to find a common denominator of a set of values is simply to multiply all the values together. The result might not be the least common denominator, but it will allow the problem to be worked.

SAMPLE QUESTIONS

21) Simplify: $2\frac{3}{5} + 3\frac{1}{4} - 1\frac{1}{2}$

Answer:

$$2\frac{3}{5} + 3\frac{1}{4} - 1\frac{1}{2}$$

$= 2\frac{12}{20} + 3\frac{5}{20} - 1\frac{10}{20}$	Change each fraction so it has a denominator of 20, which is the LCD of 5, 4, and 2.
$2 + 3 - 1 = 4$ $\frac{12}{20} + \frac{5}{20} - \frac{10}{20} = \frac{7}{20}$	Add and subtract the whole numbers together and the fractions together.
$4\frac{7}{20}$	Combine to get the final answer (a mixed number).

22) Simplify: $\frac{7}{8} \times 3\frac{1}{3}$

Answer:

$\frac{7}{8} \times 3\frac{1}{3}$	
$3\frac{1}{3} = \frac{10}{3}$	Change the mixed number to an improper fraction.
$\frac{7}{8}\left(\frac{10}{3}\right) = \frac{7 \times 10}{8 \times 3}$ $= \frac{70}{24}$	Multiply the numerators together and the denominators together.
$= \frac{35}{12}$ $= 2\frac{11}{12}$	Reduce the fraction.

23) Simplify: $4\frac{1}{2} \div \frac{2}{3}$

Answer:

$4\frac{1}{2} \div \frac{2}{3}$	
$4\frac{1}{2} = \frac{9}{2}$	Change the mixed number to an improper fraction.
$\frac{9}{2} \div \frac{2}{3}$ $= \frac{9}{2} \times \frac{3}{2}$ $= \frac{27}{4}$	Multiply the first fraction by the reciprocal of the second fraction.
$= 6\frac{3}{4}$	Simplify.

CONVERTING BETWEEN FRACTIONS AND DECIMALS

A fraction is converted to a decimal by using long division until there is no remainder and no pattern of repeating numbers occurs.

A decimal is converted to a fraction using the following steps:

▶ Place the decimal value as the numerator in a fraction with a denominator of 1.

▶ Multiply the fraction by $\frac{10}{10}$ for every digit in the decimal value, so that there is no longer a decimal in the numerator.

▶ Reduce the fraction.

SAMPLE QUESTIONS

24) Write the fraction $\frac{7}{8}$ as a decimal.

Answer:

$$
\begin{array}{r}
0.875 \\
8\overline{)7000} \\
-64 \\
\hline
60 \\
-56 \\
\hline
40
\end{array}
$$

Divide the denominator into the numerator using long division.

25) Write the fraction $\frac{5}{11}$ as a decimal.

Answer:

$$
\begin{array}{r}
0.\overline{4545} \\
11\overline{)50000} \\
-44 \\
\hline
60 \\
-55 \\
\hline
50 \\
-44 \\
\hline
60
\end{array}
$$

Dividing using long division yields a repeating decimal.

26) Write the decimal 0.125 as a fraction.

Answer:

0.125	
$= \frac{0.125}{1}$	Create a fraction with 0.125 as the numerator and 1 as the denominator.
$\frac{0.125}{1} \times \frac{10}{10} \times \frac{10}{10} \times \frac{10}{10} = \frac{125}{1000}$	Multiple by $\frac{10}{10}$ three times (one for each numeral after the decimal).
$= \frac{1}{8}$	Simplify.

Alternatively, recognize that 0.125 is read "one hundred twenty-five thousandths" and can therefore be written in fraction form as $\frac{125}{1000}$.

ROUNDING AND ESTIMATION

Rounding is a way of simplifying a complicated number. The result of rounding will be a less precise value with which it is easier to write or perform operations. Rounding is performed to a specific place value, for example the thousands or tenths place.

The rules for rounding are as follows:

1. Underline the place value being rounded to.
2. Locate the digit one place value to the right of the underlined value. If this value is less than 5, then keep the underlined value and replace all digits to the right of the underlined value with 0. If the value to the right of the underlined digit is greater than or equal to 5, then increase the underlined digit by one and replace all digits to the right of it with 0.

HELPFUL HINT

Estimation can often be used to eliminate answer choices on multiple-choice tests without having to work the problem to completion.

Estimation is when numbers are rounded and then an operation is performed. This process can be used when working with large numbers to find a close, but not exact, answer.

SAMPLE QUESTIONS

27) Round the number 138,472 to the nearest thousands.

Answer:

$138,472 \approx \mathbf{138,000}$

The 8 is in the thousands place, and the number to its right is 4. Because 4 is less than 5, the 8 remains and all numbers to the right become 0.

28) The populations of five local towns are 12,341, 8975, 9431, 10,521, and 11,427. Estimate the total population to the nearest 1000 people.

Answer:

$12,341 \approx 12,000$	
$8975 \approx 9000$	
$9431 \approx 9000$	Round each value to the thousands place.
$10,521 \approx 11,000$	
$11,427 \approx 11,000$	
$12,000 + 9000 + 9000 + 11,000 + 11,000 = \mathbf{52,000}$	Add.

FACTORIALS

A **factorial** of a number n is denoted by $n!$ and is equal to $1 \times 2 \times 3 \times 4 \times \ldots \times n$. Both $0!$ and $1!$ are equal to 1 by definition. Fractions containing factorials can often be simplified by crossing out the portions of the factorials that occur in both the numerator and denominator.

SAMPLE QUESTIONS

29) **Simplify: 8!**

Answer:

$8!$	
$= 8 \times 7 \times 6 \times 5 \times 4 \times 3 \times 2 \times 1$	Expand the factorial and multiply.
$= \mathbf{40,320}$	

30) **Simplify:** $\frac{10!}{7!3!}$

Answer:

$\frac{10!}{7!3!}$	
$= \frac{10 \times 9 \times 8 \times 7!}{7! \times 3 \times 2 \times 1}$	Expand the factorial.
$= \frac{10 \times 9 \times 8}{3 \times 2 \times 1}$	Cross out values that occur in both the numerator and denominator.
$= \frac{720}{6}$	Multiply and simplify.
$= \mathbf{120}$	

RATIOS

A **ratio** is a comparison of two numbers and can be represented as $\frac{a}{b}$, $a{:}b$, or a to b. The two numbers represent a constant relationship, not a specific value: for every a number of items in the first group, there will be b number of items in the second. For example, if the ratio of blue to red candies in a bag is 3:5, the bag will contain 3 blue candies for every 5 red candies. So, the bag might contain 3 blue candies and 5 red candies, or it might contain 30 blue candies and 50 red candies, or 36 blue candies and 60 red candies. All of these values are representative of the ratio 3:5 (which is the ratio in its lowest, or simplest, terms).

To find the "whole" when working with ratios, simply add the values in the ratio. For example, if the ratio of boys to girls in a class is 2:3, the "whole" is five: 2 out of every 5 students are boys, and 3 out of every 5 students are girls.

SAMPLE QUESTIONS

31) There are 10 boys and 12 girls in a first-grade class. What is the ratio of boys to the total number of students? What is the ratio of girls to boys?

Answer:

number of boys: 10 number of girls: 12 number of students: 22	Identify the variables.
number of boys : number of students $= 10 : 22$ $= \frac{10}{22}$ $= \frac{5}{11}$	Write out and simplify the ratio of boys to total students.
number of girls : number of boys $= 12 : 10$ $= \frac{12}{10}$ $= \frac{6}{5}$	Write out and simplify the ratio of girls to boys.

32) A family spends $600 a month on rent, $400 on utilities, $750 on groceries, and $550 on miscellaneous expenses. What is the ratio of the family's rent to their total expenses?

Answer:

rent $= 600$ utilities $= 400$ groceries $= 750$ miscellaneous $= 550$ total expenses $= 600 + 400 + 750 + 550 = 2300$	Identify the variables.
rent : total expenses $= 600 : 2300$ $= \frac{600}{2300}$ $= \frac{6}{23}$	Write out and simplify the ratio of rent to total expenses.

PROPORTIONS

A **proportion** is an equation which states that two ratios are equal. A proportion is given in the form $\frac{a}{b} = \frac{c}{d}$, where the a and d terms are the extremes and the b and *c* terms are the means. A proportion is solved using cross-multiplication ($ad = bc$) to create an equation with no fractional components. A proportion must have the same units in both numerators and both denominators.

SAMPLE QUESTIONS

33) Solve the proportion for *x*: $\frac{3x-5}{2} = \frac{x-8}{3}$.

Answer:

$\frac{(3x-5)}{2} = \frac{(x-8)}{3}$	
$3(3x-5) = 2(x-8)$	Cross-multiply.
$9x - 15 = 2x - 16$ $7x - 15 = -16$ $7x = -1$ $x = -\frac{1}{7}$	Solve the equation for *x*.

34) A map is drawn such that 2.5 inches on the map equates to an actual distance of 40 miles. If the distance measured on the map between two cities is 17.25 inches, what is the actual distance between them in miles?

Answer:

$\frac{2.5}{40} = \frac{17.25}{x}$	Write a proportion where *x* equals the actual distance and each ratio is written as inches : miles.
$2.5x = 690$ $x = 276$ The two cities are **276 miles apart**.	Cross-multiply and divide to solve for *x*.

35) A factory knows that 4 out of 1000 parts made will be defective. If in a month there are 125,000 parts made, how many of these parts will be defective?

Answer:

$\frac{4}{1000} = \frac{x}{125,000}$	Write a proportion where *x* is the number of defective parts made and both ratios are written as defective : total.
$1000x = 500,000$ $x = 500$ There are **500 defective parts** for the month.	Cross-multiply and divide to solve for *x*.

Percentages

A **percent** (or percentage) means per hundred and is expressed with a percent symbol (%). For example, 54% means 54 out of every 100. A percent can be converted to a decimal by removing the % symbol and moving the decimal point two places to the left, while a decimal can be converted to a percent by moving the decimal point two places to the right and attaching the % sign. A percent can be converted to a fraction by writing the percent as a fraction with 100 as the denominator and reducing. A fraction can be converted to a percent by performing the indicated division, multiplying the result by 100, and attaching the % sign.

The equation for finding percentages has three variables: the part, the whole, and the percent (which is expressed in the equation as a decimal). The equation, as shown below, can be rearranged to solve for any of these variables.

- ▸ part = whole × percent
- ▸ percent = $\frac{\text{part}}{\text{whole}}$
- ▸ whole = $\frac{\text{part}}{\text{percent}}$

This set of equations can be used to solve percent word problems. All that's needed is to identify the part, whole, and/or percent, and then to plug those values into the appropriate equation and solve.

SAMPLE QUESTIONS

36) **Change the following values to the indicated form:**

   A.   18% to a fraction

   B. $\frac{3}{5}$ to a percent

   C.   1.125 to a percent

   D.   84% to a decimal

   Answers:

   A.   The percent is written as a fraction over 100 and reduced: $\frac{18}{100} = \frac{9}{50}$.

   B.   Dividing 5 by 3 gives the value 0.6, which is then multiplied by 100: **60%**.

   C.   The decimal point is moved two places to the right: 1.125 × 100 = **112.5%**.

   D.   The decimal point is moved two places to the left: 84 ÷ 100 = **0.84**.

37) **In a school of 650 students, 54% of the students are boys. How many students are girls?**

Answer:

Percent of students who are girls = 100% − 54% = 46% percent = 46% = 0.46 whole = 650 students part = ?	Identify the variables.
part = whole × percent = 0.46 × 650 = 299 **There are 299 girls.**	Plug the variables into the appropriate equation.

PERCENT CHANGE

Percent change problems involve a change from an original amount. Often percent change problems appear as word problems that include discounts, growth, or markups. In order to solve percent change problems, it's necessary to identify the percent change (as a decimal), the amount of change, and the original amount. (Keep in mind that one of these will be the value being solved for.) These values can then be plugged into the equations below:

> **HELPFUL HINT**
>
> Key terms associated with percent change problems include discount, sales tax, and markup.

- ► amount of change = original amount × percent change
- ► percent change = $\dfrac{\text{amount of change}}{\text{original amount}}$
- ► original amount = $\dfrac{\text{amount of change}}{\text{percent change}}$

SAMPLE QUESTIONS

38) An HDTV that originally cost $1,500 is on sale for 45% off. What is the sale price for the item?

Answer:

original amount = $1,500 percent change = 45% = 0.45 amount of change = ?	Identify the variables.
amount of change = original amount × percent change = 1500 × 0.45 = 675	Plug the variables into the appropriate equation.
1500 − 675 = 825 **The final price is $825.**	To find the new price, subtract the amount of change from the original price.

39) A house was bought in 2000 for $100,000 and sold in 2015 for $120,000. What was the percent growth in the value of the house from 2000 to 2015?

Answer:

original amount = $100,000 amount of change = 120,000 − 100,000 = 20,000 percent change = ?	Identify the variables.
percent change = $\frac{\text{amount of change}}{\text{original amount}}$ $= \frac{20,000}{100,000}$ $= 0.20$	Plug the variables into the appropriate equation.
$0.20 \times 100 = $ **20%**	To find the percent growth, multiply by 100.

COMPARISON OF RATIONAL NUMBERS

Rational numbers can be ordered from least to greatest (or greatest to least) by placing them in the order in which they fall on a number line. When comparing a set of fractions, it's often easiest to convert each value to a common denominator. Then, it's only necessary to compare the numerators of each fraction.

When working with numbers in multiple forms (for example, a group of fractions and decimals), convert the values so that the set contains only fractions or only decimals. When ordering negative numbers, remember that the negative numbers with the largest absolute values are farthest from 0 and are therefore the smallest numbers. (For example, –75 is smaller than –25.)

HELPFUL HINT

Drawing a number line can help when comparing numbers: the final list should go in order from left to right (least to greatest) or right to left (greatest to least) on the line.

SAMPLE QUESTIONS

40) Order the following numbers from greatest to least: $-\frac{2}{3}$, 1.2, 0, –2.1, $\frac{5}{4}$, –1, $\frac{1}{8}$.

Answer:

$-\frac{2}{3} = -0.\overline{66}$ $\frac{5}{4} = 1.25$ $\frac{1}{8} = 0.125$	Change each fraction to a decimal.

1.25, 1.2, 0.125, 0, $-0.\overline{66}$, -1, -2.1	Place the decimals in order from greatest to least.
$\frac{5}{4}$, 1.2, $\frac{1}{8}$, 0, $-\frac{2}{3}$, -1, -2.1	Convert back to fractions if the problem requires it.

41) Order the following numbers from least to greatest: $\frac{1}{3}$, $-\frac{5}{6}$, $1\frac{1}{8}$, $\frac{7}{12}$, $-\frac{3}{4}$, $-\frac{3}{2}$.

Answer:

$\frac{1}{3} = \frac{8}{24}$ $-\frac{5}{6} = -\frac{20}{24}$ $1\frac{1}{8} = \frac{9}{8} = \frac{27}{24}$ $\frac{7}{12} = \frac{14}{24}$ $-\frac{3}{4} = -\frac{18}{24}$ $-\frac{3}{2} = -\frac{36}{24}$	Convert each value using the least common denominator of 24.
$-\frac{36}{24}$, $-\frac{20}{24}$, $-\frac{18}{24}$, $\frac{8}{24}$, $\frac{14}{24}$, $\frac{27}{24}$	Arrange the fractions in order from least to greatest by comparing the numerators.
$-\frac{3}{2}$, $-\frac{5}{6}$, $-\frac{3}{4}$, $\frac{1}{3}$, $\frac{7}{12}$, $1\frac{1}{8}$	Put the fractions back in their original form if the problem requires it.

EXPONENTS AND RADICALS

EXPONENTS

An expression in the form b^n is in an exponential notation where b is the **base** and n is an **exponent**. To perform the operation, multiply the base by itself the number of times indicated by the exponent. For example, 2^3 is equal to $2 \times 2 \times 2$ or 8.

Table 1.6. Operations with Exponents

Rule	Example	Explanation
$a^0 = 1$	$5^0 = 1$	Any base (except 0) to the 0 power is 1.
$a^{-n} = \frac{1}{a^n}$	$5^{-3} = \frac{1}{5^3}$	A negative exponent becomes positive when moved from numerator to denominator (or vice versa).
$a^m a^n = a^{m+n}$	$5^3 5^4 = 5^{3+4} = 5^7$	Add the exponents to multiply two powers with the same base.

Table 1.6. Operations with Exponents (continued)

Rule	Example	Explanation
$(a^m)^n = a^{m \times n}$	$(5^3)^4 = 5^{3(4)} = 5^{12}$	Multiply the exponents to raise a power to a power.
$\dfrac{a^m}{a^n} = a^{m-n}$	$\dfrac{5^4}{5^3} = 5^{4-3} = 5^1$	Subtract the exponents to divide two powers with the same base.
$(ab)^n = a^n b^n$	$(5 \times 6)^3 = 5^3 6^3$	Apply the exponent to each base to raise a product to a power.
$\left(\dfrac{a}{b}\right)^n = \dfrac{a^n}{b^n}$	$\left(\dfrac{5}{6}\right)^3 = \dfrac{5^3}{6^3}$	Apply the exponent to each base to raise a quotient to a power.
$\left(\dfrac{a}{b}\right)^{-n} = \left(\dfrac{b}{a}\right)^n$	$\left(\dfrac{5}{6}\right)^{-3} = \left(\dfrac{6}{5}\right)^3$	Invert the fraction and change the sign of the exponent to raise a fraction to a negative power.
$\dfrac{a^m}{b^n} = \dfrac{b^{-n}}{a^{-m}}$	$\dfrac{5^3}{6^4} = \dfrac{6^{-4}}{5^{-3}}$	Change the sign of the exponent when moving a number from the numerator to denominator (or vice versa).

SAMPLE QUESTIONS

42) **Simplify:** $\dfrac{(10^2)^3}{(10^2)^2}$

Answer:

$\dfrac{(10^2)^3}{(10^2)^2}$	
$= \dfrac{10^6}{10^4}$	Multiply the exponents raised to a power.
$= 10^{6-4}$	Subtract the exponent in the denominator from the one in the numerator.
$= 10^2$ $= \mathbf{100}$	Simplify.

43) **Simplify:** $\dfrac{(x^{-2}y^2)^2}{x^3 y}$

Answer:

$\dfrac{(x^{-2}y^2)^2}{x^3 y}$	
$= \dfrac{x^{-4}y^4}{x^3 y}$	Multiply the exponents raised to a power.

$= x^{-4-3}y^{4-1}$ $= x^{-7}y^3$	Subtract the exponent in the denominator from the one in the numerator.
$= \dfrac{y^3}{x^7}$	Move negative exponents to the denominator.

RADICALS

Radicals are expressed as $\sqrt[b]{a}$, where b is called the **index** and a is the **radicand**. A radical is used to indicate the inverse operation of an exponent: finding the base which can be raised to b to yield a. For example, $\sqrt[3]{125}$ is equal to 5 because $5 \times 5 \times 5$ equals 125. The same operation can be expressed using a fraction exponent, so $\sqrt[b]{a} = a^{\frac{1}{b}}$. Note that when no value is indicated for b, it is assumed to be 2 (square root).

When b is even and a is positive, $\sqrt[b]{a}$ is defined to be the positive real value n such that $n^b = a$ (example: $\sqrt{16} = 4$ only, and not –4, even though $(-4)(-4) = 16$). If b is even and a is negative, $\sqrt[b]{a}$ will be a complex number (example: $\sqrt{-9} = 3i$). Finally if b is odd, $\sqrt[b]{a}$ will always be a real number regardless of the sign of a. If a is negative, $\sqrt[b]{a}$ will be negative since a number to an odd power is negative (example: $\sqrt[5]{-32} = -2$ since $(-2)^5 = -32$).

$\sqrt[n]{x}$ is referred to as the nth root of x.

▸ $n = 2$ is the square root

▸ $n = 3$ is the cube root

▸ $n = 4$ is the fourth root

▸ $n = 5$ is the fifth root

The following table of operations with radicals holds for all cases EXCEPT the case where b is even and a is negative (the complex case).

Table 1.7. Operations with Radicals

Rule	Example	Explanation
$\sqrt[b]{ac} = \sqrt[b]{a}\,\sqrt[b]{c}$	$\sqrt[3]{81} = \sqrt[3]{27}\,\sqrt[3]{3} = 3\sqrt[3]{3}$	The values under the radical sign can be separated into values that multiply to the original value.
$\sqrt[b]{\dfrac{a}{c}} = \dfrac{\sqrt[b]{a}}{\sqrt[b]{c}}$	$\sqrt{\dfrac{4}{81}} = \dfrac{\sqrt{4}}{\sqrt{81}} = \dfrac{2}{9}$	The b-root of the numerator and denominator can be calculated when there is a fraction under a radical sign.
$\sqrt[b]{a^c} = (\sqrt[b]{a})^c = a^{\frac{c}{b}}$	$\sqrt[3]{6^2} = (\sqrt[3]{6})^2 = 6^{\frac{2}{3}}$	The b-root can be written as a fractional exponent. If there is a power under the radical sign, it will be the numerator of the fraction.

Table 1.7. Operations with Radicals (continued)

Rule	Example	Explanation
$\dfrac{c}{\sqrt[b]{a}} \times \dfrac{\sqrt[b]{a}}{\sqrt[b]{a}} = \dfrac{c\sqrt[b]{a}}{a}$	$\dfrac{5}{\sqrt{2}}\dfrac{\sqrt{2}}{\sqrt{2}} = \dfrac{5\sqrt{2}}{2}$	To rationalize the denominator, multiply the numerator and denominator by the radical in the denominator until the radical has been canceled out.
$\dfrac{c}{b-\sqrt{a}} \times \dfrac{b+\sqrt{a}}{b+\sqrt{a}}$ $= \dfrac{c(b+\sqrt{a})}{b^2-a}$	$\dfrac{4}{3-\sqrt{2}}\dfrac{3+\sqrt{2}}{3+\sqrt{2}}$ $= \dfrac{4(3+\sqrt{2})}{9-2} = \dfrac{12+4\sqrt{2}}{7}$	To rationalize the denominator, the numerator and denominator are multiplied by the conjugate of the denominator.

SAMPLE QUESTIONS

44) **Simplify:** $\sqrt{48}$

Answer:

$\sqrt{48}$	
$= \sqrt{16 \times 3}$	Determine the largest square number that is a factor of the radicand (48) and write the radicand as a product using that square number as a factor.
$= \sqrt{16}\sqrt{3}$ $= \mathbf{4\sqrt{3}}$	Apply the rules of radicals to simplify.

45) **Simplify:** $\dfrac{6}{\sqrt{8}}$

Answer:

$\dfrac{6}{\sqrt{8}}$	
$= \dfrac{6}{\sqrt{4}\sqrt{2}}$ $= \dfrac{6}{2\sqrt{2}}$	Apply the rules of radicals to simplify.
$= \dfrac{6}{2\sqrt{2}}\left(\dfrac{\sqrt{2}}{\sqrt{2}}\right)$ $= \dfrac{\mathbf{3\sqrt{2}}}{\mathbf{2}}$	Multiply by $\dfrac{\sqrt{2}}{\sqrt{2}}$ to rationalize the denominator.

MATRICES

A **matrix** is a rectangular arrangement of numbers into **rows** (horizontal set of numbers) and **columns** (vertical set of numbers). A matrix with the same number of rows and columns is called a **square matrix**.

The **dimensions** of a matrix are given as $m \times n$, where m is the number of rows and n is the number of columns.

$$A = \begin{bmatrix} 1 & 8 & -2 \\ -12 & -3 & 7 \end{bmatrix} \leftarrow \text{row}$$

matrix name   column

Figure 1.5. Parts of a Matrix

MATRIX OPERATIONS

Matrices can be added and subtracted together if and only if the matrices have the same dimensions. When the matrices have the same dimensions, the values in corresponding positions in the matrices can be added or subtracted. For example, the elements in row 1 column 1 from each matrix are added or subtracted. The matrix operation of addition is both commutative and associative: $A + B = B + A$ and $(A + B) + C = A + (B + C)$.

Matrices can be multiplied by a single value called a scalar. To perform this operation, each value in the matrix is multiplied by the scalar.

Two matrices can be multiplied only when the number of columns in the first matrix (with dimensions $m \times n$) equals the number of rows in the second matrix (dimensions $n \times p$). The resulting matrix will have dimensions $m \times p$. To complete the multiplication, the values in each row are multiplied by the corresponding values

> **HELPFUL HINT**
>
> $$\begin{bmatrix} a & b \\ c & d \end{bmatrix} \pm \begin{bmatrix} e & f \\ g & h \end{bmatrix} = \begin{bmatrix} a \pm e & b \pm f \\ c \pm g & d \pm h \end{bmatrix}$$
>
> $$x \begin{bmatrix} a & b \\ c & d \end{bmatrix} = \begin{bmatrix} xa & xb \\ xc & xd \end{bmatrix}$$

in each column (e.g., the first value in the row by the first value in the column, the second value in the row by the second value in the row, and so forth). The results of each row by column multiplication are then added together to give a single value. This resulting sum of products of row a in the first matrix by column b in the second matrix is placed in row a, column b of the resulting matrix.

The matrix operation of multiplication is NOT commutative ($AB \neq BA$), but is associative ($A(BC) = (AB)C$). The distributive property of multiplication over addition also holds for matrices ($(A(B + C) = AB + BC$).

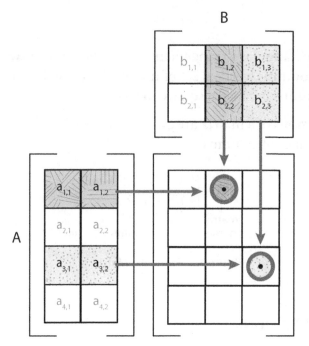

Figure 1.6. Matrix Multiplication

SAMPLE QUESTIONS

46) Simplify: $\begin{bmatrix} 7 & 4 & 1 \\ -6 & 3 & 5 \end{bmatrix} + \begin{bmatrix} 2 & -8 & 5 \\ -1 & 0 & 4 \end{bmatrix}$

Answer:

$\begin{bmatrix} 7 & 4 & 1 \\ -6 & 3 & 5 \end{bmatrix} + \begin{bmatrix} 2 & -8 & 5 \\ -1 & 0 & 4 \end{bmatrix}$	
$\begin{bmatrix} 7+2 & 4+(-8) & 1+5 \\ -6+(-1) & 3+0 & 5+4 \end{bmatrix}$ $= \begin{bmatrix} \mathbf{9} & \mathbf{-4} & \mathbf{6} \\ \mathbf{-7} & \mathbf{3} & \mathbf{9} \end{bmatrix}$	Add the corresponding elements in each matrix.

47) Simplify: $\frac{1}{3} \begin{bmatrix} 10 & 15 & -4 \\ 9 & 20 & 16 \\ 0 & 14 & -3 \end{bmatrix}$

Answer:

$\frac{1}{3} \begin{bmatrix} 10 & 15 & -4 \\ 9 & 20 & 16 \\ 0 & 14 & -3 \end{bmatrix}$	

$$= \begin{bmatrix} \frac{1}{3}(10) & \frac{1}{3}(15) & \frac{1}{3}(-4) \\ \frac{1}{3}(9) & \frac{1}{3}(20) & \frac{1}{3}(16) \\ \frac{1}{3}(0) & \frac{1}{3}(14) & \frac{1}{3}(-3) \end{bmatrix}$$

Multiply each term in the matrix by $\frac{1}{3}$.

$$= \begin{bmatrix} \frac{10}{3} & 5 & -\frac{4}{3} \\ 3 & \frac{20}{3} & \frac{16}{3} \\ 0 & \frac{14}{3} & -1 \end{bmatrix}$$

48) **Simplify:** $\begin{bmatrix} 1 & 0 & -2 \\ -3 & 4 & 1 \end{bmatrix} \begin{bmatrix} 3 & 0 \\ -2 & 4 \\ 1 & -4 \end{bmatrix}$

Answer:

$$\begin{bmatrix} 1 & 0 & -2 \\ -3 & 4 & 1 \end{bmatrix} \begin{bmatrix} 3 & 0 \\ -2 & 4 \\ 1 & -4 \end{bmatrix}$$

The matrices can be multiplied together because the number of columns in the first matrix equals the number of rows in the second matrix.

$$= \begin{bmatrix} 1(3) + 0(-2) - 2(1) & 1(0) + 0(4) - 2(-4) \\ -3(3) + 4(-2) + 1(1) & -3(0) + 4(4) + 1(-4) \end{bmatrix}$$

$$= \begin{bmatrix} 3 + 0 - 2 & 0 + 0 + 8 \\ -9 - 8 + 1 & 0 + 16 - 4 \end{bmatrix}$$

$$= \begin{bmatrix} 1 & 8 \\ -16 & 12 \end{bmatrix}$$

Multiply the values in each row of the first matrix by the values in each column of the second and add. The resulting matrix is 2×2.

DETERMINANTS

The **determinant** of a matrix (written as det(*A*) or |*A*|) is a value calculated by manipulating elements of a square matrix. The determinant of a 2 × 2 or a 3 × 3 matrix can easily be found by hand, but determinants of larger matrices are usually found using a calculator.

$$\begin{vmatrix} a & b \\ c & d \end{vmatrix} = ad - bc$$

$$\begin{vmatrix} a & b & c \\ d & e & f \\ g & h & i \end{vmatrix} = a \begin{vmatrix} e & f \\ h & i \end{vmatrix} - b \begin{vmatrix} d & f \\ g & i \end{vmatrix} + c \begin{vmatrix} d & e \\ g & h \end{vmatrix}$$

$$= a(ei - fh) - b(di - fg) + c(dh - eg)$$

49) **Find the determinant of the matrix** $\begin{bmatrix} 7 & -3 \\ 4 & -2 \end{bmatrix}$**.**

Answer:

$\begin{vmatrix} 7 & -3 \\ 4 & -2 \end{vmatrix}$

$= 7(-2) - (-3)(4)$

$= -14 + 12$

$= \mathbf{-2}$

Use the formula to find the determinant of a 2 × 2 matrix.

50) **Find the determinant of the matrix** $\begin{bmatrix} -2 & 4 & 1 \\ 0 & 3 & -5 \\ 7 & -1 & 4 \end{bmatrix}$**.**

Answer:

$\begin{bmatrix} -2 & 4 & 1 \\ 0 & 3 & -5 \\ 7 & -1 & 4 \end{bmatrix}$

$\begin{vmatrix} -2 & 4 & 1 \\ 0 & 3 & -5 \\ 7 & -1 & 4 \end{vmatrix}$

$= -2\begin{vmatrix} 3 & -5 \\ -1 & 4 \end{vmatrix} - 4\begin{vmatrix} 0 & -5 \\ 7 & 4 \end{vmatrix} + 1\begin{vmatrix} 0 & 3 \\ 7 & -1 \end{vmatrix}$

$= -2(3(4) - (-5)(-1)) - 4(0(4) - (-5)7) + 1(0(-1) - 7(3))$

$= -2(12 - 5) - 4(0 + 35) + 1(0 - 21)$

$= -2(7) - 4(35) + 1(-21)$

$= -14 - 140 - 21$

$= \mathbf{-175}$

Use the formula to find the determinant of a 3 × 3 matrix.

IDENTITY AND INVERSE MATRICES

The **identity matrix** (**I**) is a square matrix with values of 1 forming a diagonal from the upper left corner to the bottom right corner; the rest of the elements are 0. When performing multiplication, the identity matrix functions like the number 1: the product of a matrix *A* and the identity matrix *I* returns the original matrix *A* ($A \times I = A$).

$$\begin{bmatrix} 1 & 0 & 0 \\ 0 & 1 & 0 \\ 0 & 0 & 1 \end{bmatrix}$$

Figure 1.7. Identity Matrix

An **inverse matrix** (A^{-1}) is a square matrix that, when multiplied by the original matrix, results in the identity matrix ($A \times A^{-1} = I$). An inverse matrix can only be calculated for square matrices. A matrix is not invertible if its determinant is 0 (since the formula requires dividing by the determinant and division by 0 is not defined).

$$\begin{bmatrix} a & b \\ c & d \end{bmatrix}^{-1} = \frac{1}{ad-bc}\begin{bmatrix} d & -b \\ -c & a \end{bmatrix}$$

$$\begin{bmatrix} a & b & c \\ d & e & f \\ g & h & i \end{bmatrix}^{-1} = \frac{1}{\det A}\begin{bmatrix} ei-fh & ch-bi & bf-ce \\ fg-di & ai-cg & cd-af \\ dh-eg & bg-ah & ae-bd \end{bmatrix}$$

Multiplying a matrix by its inverse functions like division for matrices. (Note that there is no "real" division for matrices.) This operation can be used to solve matrix equations by setting up a system of equations using matrices (as discussed in "Matrix Operations").

SAMPLE QUESTIONS

51) If $A = \begin{bmatrix} 2 & 3 \\ -1 & 4 \end{bmatrix}$, find A^{-1}.

Answer:

$$\begin{bmatrix} 2 & 3 \\ -1 & 4 \end{bmatrix}^{-1}$$

$$= \frac{1}{11}\begin{bmatrix} 4 & -3 \\ 1 & 2 \end{bmatrix}$$

$$= \begin{bmatrix} \frac{4}{11} & \frac{-3}{11} \\ \frac{1}{11} & \frac{2}{11} \end{bmatrix}$$

Use the formula to find the inverse of a 2 × 2 matrix.

Check the answer: $A \times A^{-1} = \begin{bmatrix} 2 & 3 \\ -1 & 4 \end{bmatrix}\begin{bmatrix} \frac{4}{11} & \frac{-3}{11} \\ \frac{1}{11} & \frac{2}{11} \end{bmatrix} = \begin{bmatrix} 1 & 0 \\ 0 & 1 \end{bmatrix}$

52) **Find the inverse matrix of** $\begin{bmatrix} 1 & -1 & 2 \\ -1 & 0 & -2 \\ 2 & 1 & -2 \end{bmatrix}$.

Answer:

$$\begin{bmatrix} 1 & -1 & 2 \\ -1 & 0 & -2 \\ 2 & 1 & -2 \end{bmatrix}^{-1}$$

$$= \frac{1}{6}\begin{bmatrix} 0(-2)-(-2)1 & 2(1)-(-1)(-2) & (-1)(-2)-2(0) \\ (-2)(2)-(-1)(-2) & 1(-2)-2(2) & 2(-1)-1(-2) \\ (-1)(1)-0(2) & (-1)2-1(1) & 1(0)-(-1)(-1) \end{bmatrix}$$

Use the formula to find the inverse of a 3 × 3 matrix.

$$= \frac{1}{6}\begin{bmatrix} 2 & 0 & 2 \\ -6 & -6 & 0 \\ -1 & -3 & -1 \end{bmatrix}$$

$$= \begin{bmatrix} \frac{1}{3} & 0 & \frac{1}{3} \\ -1 & -1 & 0 \\ -\frac{1}{6} & -\frac{1}{2} & -\frac{1}{6} \end{bmatrix}$$

Use the formula to find the inverse of a 3 × 3 matrix.

TRANSFORMATIONS IN THE PLANE WITH MATRICES

An important application of matrix operations is transforming figures in a plane.

A matrix can represent a polygon in a plane. Each vertex (x,y) of the polygon can be stored in a matrix, with all the x values in row 1 and all the y values in row 2. For example, the triangle with vertices at (2,1), (6,5) and (4,–3) would be represented by matrix $\begin{bmatrix} 2 & 6 & 4 \\ 1 & 5 & -3 \end{bmatrix}$. Transformations of the polygon can then be represented by matrix operations.

Horizontal and vertical translations can be represented by adding a matrix of the same dimensions of the polygon matrix to that matrix. For example, adding the matrix $\begin{bmatrix} 3 & 3 & 3 \\ -2 & -2 & -2 \end{bmatrix}$ to the matrix representing the triangle above would result in a matrix that represents a congruent triangle that has been horizontally translated 3 units to the right and vertically translated 2 units down.

Dilations and compressions can be represented by multiplying a polygon matrix by a square matrix in the form $\begin{bmatrix} k & 0 \\ 0 & 1 \end{bmatrix}$ for horizontal dilations ($k > 1$) or compressions ($0 < k < 1$) or $\begin{bmatrix} 1 & 0 \\ 0 & k \end{bmatrix}$ for vertical dilations or compressions. Notice that these matrices resemble the identity matrix, with only one value changed. This allows for only x values or y values to be adjusted as desired.

A reflection across the y axis can be achieved by changing the signs of all the x values of the original polygon by multiplying be a square matrix $\begin{bmatrix} -1 & 0 \\ 0 & 1 \end{bmatrix}$. Similarly, a reflection across the x axis is achieved by changing the signs of the y values by multiplying by the matrix $\begin{bmatrix} 1 & 0 \\ 0 & -1 \end{bmatrix}$.

Finally, rotations can also be represented by multiplication by square matrices:

► counterclockwise 90°: $\begin{bmatrix} 0 & -1 \\ 1 & 0 \end{bmatrix}$

► counterclockwise 180°: $\begin{bmatrix} -1 & 0 \\ 0 & -1 \end{bmatrix}$

► counterclockwise 270° (clockwise 90°): $\begin{bmatrix} 0 & 1 \\ -1 & 0 \end{bmatrix}$

Matrices can also be used to find the area of a triangle with vertices (x_1, y_1), (x_2, y_2), and (x_3, y_3) using the formula below. If A is negative, take the absolute value to make it positive; this is the area.

$$A = \frac{1}{2} \begin{vmatrix} x_1 & y_1 & 1 \\ x_2 & y_2 & 1 \\ x_3 & y_3 & 1 \end{vmatrix}$$

SAMPLE QUESTION

53) Find the new vertices of a triangle with vertices $(2, 3)$, $(2, 7)$, $(6, 3)$ after it has been rotated clockwise by $90°$ and then compressed vertically by a factor of $\frac{1}{2}$. Then find the area of the resultant triangle.

Answer:

$(2, 3), (2, 7), (6, 3) \rightarrow \begin{bmatrix} 2 & 2 & 6 \\ 3 & 7 & 3 \end{bmatrix}$	Set up a matrix using the vertices of the triangle.
$\begin{bmatrix} 0 & 1 \\ -1 & 0 \end{bmatrix} \begin{bmatrix} 2 & 2 & 6 \\ 3 & 7 & 3 \end{bmatrix}$ $= \begin{bmatrix} 3 & 7 & 3 \\ -2 & -2 & -6 \end{bmatrix}$	Multiple by the appropriate matrix to rotate the triangle clockwise $90°$.
$\begin{bmatrix} 1 & 0 \\ 0 & \frac{1}{2} \end{bmatrix} \begin{bmatrix} 3 & 7 & 3 \\ -2 & -2 & -6 \end{bmatrix}$ $= \begin{bmatrix} 3 & 7 & 3 \\ -1 & -1 & -3 \end{bmatrix}$	Multiple by the appropriate matrix to compress the rotated triangle.
$A = \frac{1}{2} \begin{vmatrix} 3 & -1 & 1 \\ 7 & -1 & 1 \\ 3 & -3 & 1 \end{vmatrix}$ $= \frac{1}{2}(-8)$ $= -4$ **The area of the triangle is 4 units.**	Use the formula to find the area of the new triangle using the determinant.

SEQUENCES AND SERIES

Sequences can be thought of as a set of numbers (called **terms**) with a rule that explains the particular pattern between the terms. The terms of a sequence are separated by commas. There are two types of sequences that will be examined, arithmetic and geometric. The sum of an arithmetic sequence is known as an **arithmetic series**; similarly the sum of a geometric sequence is known as a **geometric series**.

ARITHMETIC SEQUENCES

Arithmetic growth is constant growth, meaning that the difference between any one term in the series and the next consecutive term will be the same constant. This constant is called the **common difference**. Thus, to list the terms in the sequence, one can just add (or subtract) the same number repeatedly. For example, the series {20, 30, 40, 50} is arithmetic since 10 is added each time to get from one term to the next. One way to represent this sequence is using a **recursive** definition, which basically says: *next term = current term + common difference*. For this example, the recursive definition would be $a_{n+1} = a_n + 10$ because the *next* term a_{n+1} in the sequence is the current term a_n plus 10. In general, the recursive definition of a series is:

$$a_{n+1} = a_n + d,\text{ where } d \text{ is the common difference.}$$

Often, the objective of arithmetic sequence questions is to find a specific term in the sequence or the sum of a certain series of terms. The formulas to use are:

Table 1.8. Formulas for Arithmetic Sequences and Series

Finding the *n*th term . . .	
$a_n = a_1 + d(n-1)$ $a_n = a_m + d(n-m)$	d = the common difference of the sequence a_n = the *n*th term in the sequence n = the number of the term a_m = the *m*th term in the sequence m = the number of the term a_1 = the first term in the sequence
Finding the partial sum . . .	
$S_n = \dfrac{n(a_1 + a_n)}{2}$	S_n = sum of the terms through the *n*th term a_n = the *n*th term in the sequence n = the number of the term a_1 = the first term in the sequence

SAMPLE QUESTIONS

54) **Find the ninth term of the sequence: −57, −40, −23, −6 . . .**

Answer:

$a_1 = -57$ $d = -57 - (-40) = 17$ $n = 9$	Identify the variables given.
$a_9 = -57 + 17(9 - 1)$	Plug these values into the formula for the specific term of an arithmetic sequence.

$a_9 = -57 + 17(8)$	
$a_9 = -57 + 136$	Solve for a_9.
$\mathbf{a_9 = 79}$	

55) **If the 23rd term in an arithmetic sequence is 820, and the 5th term is 200, find the common difference between each term.**

Answer:

$a_5 = 200$	
$a_{23} = 820$	
$n = 23$	Idenfity the variables given.
$m = 5$	
$d = ?$	
$a_n = a_m + d(n - m)$	
$820 = 200 + d(23 - 5)$	Plug these values into the equation for using one term to find another in an arithmetic sequence.
$620 = d(18)$	
$\mathbf{d = 34.\overline{44}}$	

56) **Evaluate $\sum_{n=14}^{45} 2n + 10$.**

Answer:

$a_1 = 2(1) + 10 = 12$	
$n = 45$	
$a_n = 2(45) + 10 = 100$	
$S_n = \dfrac{n(a_1 + a_n)}{2}$	Find the partial sum of the first 45 terms.
$= \dfrac{45(12 + 100)}{2}$	
$= 2520$	
$a_1 = 2(1) + 10 = 12$	
$n = 13$	
$a_n = 2(13) + 10 = 36$	
$S_n = \dfrac{n(a_1 + a_n)}{2}$	Find the partial sum of the first 13 terms.
$= \dfrac{13(12 + 36)}{2}$	
$= 312$	
$S_{45} - S_{13} = 2520 - 312$	The sum of the terms between 14 and 45 will be the difference between S_{45} and S_{13}.
$= \mathbf{2208}$	

GEOMETRIC SEQUENCES

While an arithmetic sequence has an additive pattern, a **geometric sequence** has a multiplicative pattern. This means that to get from any one term in the sequence to the next term in the sequence, the term is multiplied by a fixed number (called the **common ratio**). The following sequence is a geometric sequence: {8, 4, 2, 1, .5, .25, .125}. In this case, the multiplier (or common ratio) is $\frac{1}{2}$. The multiplier can be any real number other than 0 or 1. To find the common ratio, simply choose any term in the sequence and divide it by the previous term (this is the ratio of two consecutive terms—thus the name common *ratio*). In the above example, the ratio between the second and third terms is $\frac{2}{4} = \frac{1}{2}$.

> **QUICK REVIEW**
>
> Compared to arithmetic growth, geometric growth is much faster. As seen in the formulas used to find a geometric term, geometric growth is exponential, whereas arithmetic growth is linear.

Geometric sequences require their own formulas to find the next term and a sum of a specific series.

Table 1.9. Geometric Sequences: Formulas

Finding the nth term . . .	
$a_n = a_1 \times r^{n-1}$ $a_n = a_m \times r^{n-m}$	r = the common ratio of the sequence a_n = the nth term in the sequence n = the number of the term a_m = the mth term in the sequence m = the number of the term a_1 = the first term in the sequence
Finding the partial sum . . .	
$S_n = \dfrac{a_1(1 - r^n)}{1 - r}$	S_n = sum of the terms through the nth term r = the common ratio of the sequence a_n = the nth term in the sequence n = the number of the term a_1 = the first term in the sequence
Finding the sum of an infinite series . . .	
$S_\infty = \dfrac{a}{1 - r}$ ($|r| < 1$)	S_∞ = sum of all terms r = the common ratio of the sequence a = the fifth term in the sequence

The finite sum formula works similarly to the arithmetic sequence sum. However, sometimes the **infinite sum** of the sequence must be found. The sum of an infinite number of terms of a sequence is called a **series**. If the infinite terms of the sequence add up to a finite number, the series is said to **converge** to that number. If the sum of the terms is infinite, then the series **diverges**. Another way to say this is to ask: is there a limit to the finite sum S_n as n goes to infinity? For geometric series in the form $\sum_{n=1}^{\infty} a \times r^n$, the series converges only when $|r| < 1$ (or $-1 < r < 1$). If r is greater than 1, the sum will approach infinity, so the series diverges.

SAMPLE QUESTIONS

57) **Find the 8th term in the sequence: {13, 39, 117, 351 . . .}**

Answer:

$a_1 = 13$	
$n = 8$	Identify the variables given.
$r = \frac{39}{13} = 3$	
$a_8 = 13 \times 3^{8-1}$	Plug these values into the equation to find
$a_8 = 13 \times 2187 = 28{,}431$	a specific term in a geometric sequence.
The eighth term of the given sequence is **28,431**.	

58) **Find the sum of the first 10 terms of this sequence: {−4, 16, −64, 256 . . .}**

Answer:

$a_1 = -4$	
$n = 10$	Identify the variables given.
$r = \frac{16}{-4} = -4$	
$S_{10} = \dfrac{-4(1 - (-4)^{10})}{1 - (-4)}$	
$= \dfrac{-4(1 - 1{,}048{,}576)}{5}$	Plug these values into the equation for the partial sum of a geometric sequence.
$= \dfrac{4{,}194{,}300}{5}$	
$= \mathbf{838{,}860}$	

2

Algebra

Algebra, meaning "restoration" in Arabic, is the mathematical method of finding the unknown. The first algebraic book in Egypt was used to figure out complex inheritances that were to be split among many individuals. Today, algebra is just as necessary when dealing with unknown amounts.

ALGEBRAIC EXPRESSIONS

The foundation of algebra is the **variable**, an unknown number represented by a symbol (usually a letter such as x or a). Variables can be preceded by a **coefficient**, which is a constant (i.e., a real number) in front of the variable, such as $4x$ or $-2a$. An **algebraic expression** is any sum, difference, product, or quotient of variables and numbers (for example $3x^2$, $2x + 7y - 1$, and $\frac{5}{x}$ are algebraic expressions). **Terms** are any quantities that are added or subtracted (for example, the terms of the expression $x^2 - 3x + 5$ are x^2, $3x$, and 5). A **polynomial expression** is an algebraic expression where all the exponents on the variables are whole numbers. A polynomial with only two terms is known as a **binomial**, and one with three terms is a **trinomial**. A **monomial** has only one term.

coefficient

$$x^2 + 3y - 12$$

variable   constant

Figure 2.1. Polynomial Expression

Evaluating expressions is another way of saying "find the numeric value of an expression if the variable is equal to a certain number." To evaluate the expression, simply plug the given value(s) for the variable(s) into the equation and simplify. Remember to use the order of operations when simplifying:

1.	Parentheses	4.	Division
2.	Exponents	5.	Addition
3.	Multiplication	6.	Subtraction

SAMPLE QUESTION

1)   If $m = 4$, find the value of the following expression:
$5(m - 2)^3 + 3m^2 - \frac{m}{4} - 1$

Answer:

$5(m - 2)^3 + 3m^2 - \frac{m}{4} - 1$	
$= 5(4 - 2)^3 + 3(4)^2 - \frac{4}{4} - 1$	Plug the value 4 in for m in the expression.
$= 5(2)^3 + 3(4)^2 - \frac{4}{4} - 1$	Calculate all the expressions inside the parentheses.
$= 5(8) + 3(16) - \frac{4}{4} - 1$	Simplify all exponents.
$= 40 + 48 - 1 - 1$	Perform multiplication and division from left to right.
$= \mathbf{86}$	Perform addition and subtraction from left to right.

OPERATIONS WITH EXPRESSIONS

ADDING AND SUBTRACTING

Expressions can be added or subtracted by simply adding and subtracting **like terms**, which are terms with the same variable part (the variables must be the same, with the same exponents on each variable). For example, in the expressions $2x + 3xy - 2z$ and $6y + 2xy$, the like terms are $3xy$ and $2xy$. Adding the two expressions yields the new expression $2x + 6xy - 2z + 6y$. Note that the other terms did not change; they cannot be combined because they have different variables.

SAMPLE QUESTION

2)   If $a = 12x + 7xy - 9y$ and $b = 8x - 9xz + 7z$, what is $a + b$?

Answer:

$a + b = (12x + 8x) + 7xy - 9y - 9xz + 7z$ $= \mathbf{20x + 7xy - 9y - 9xz + 7z}$	The only like terms in both expressions are 12x and 8x, so these two terms will be added, and all other terms will remain the same.

DISTRIBUTING AND FACTORING

Distributing and factoring can be seen as two sides of the same coin. **Distribution** multiplies each term in the first factor by each term in the second factor to get rid of parentheses. **Factoring** reverses this process, taking a polynomial in standard form and writing it as a product of two or more factors.

When distributing a monomial through a polynomial, the expression outside the parentheses is multiplied by each term inside the parentheses. Using the rules of exponents, coefficients are multiplied and exponents are added.

> **HELPFUL HINT**
>
> Operations with polynomials can always be checked by evaluating equivalent expressions for the same value.

When simplifying two polynomials, each term in the first polynomial must multiply each term in the second polynomial. A binomial (two terms) multiplied by a binomial, will require 2×2 or 4 multiplications. For the binomial × binomial case, this process is sometimes called **FOIL**, which stands for first, outside, inside, and last. These terms refer to the placement of each term of the expression: multiply the first term in each expression, then the outside terms, then the inside terms, and finally the last terms. A binomial (two terms) multiplied by a trinomial (three terms), will require 2×3 or 6 products to simplify. The first term in the first polynomial multiplies each of the three terms in the second polynomial, then the second term in the first polynomial multiplies each of the three terms in the second polynomial. A trinomial (three terms) by a trinomial will require 3×3 or 9 products, and so on.

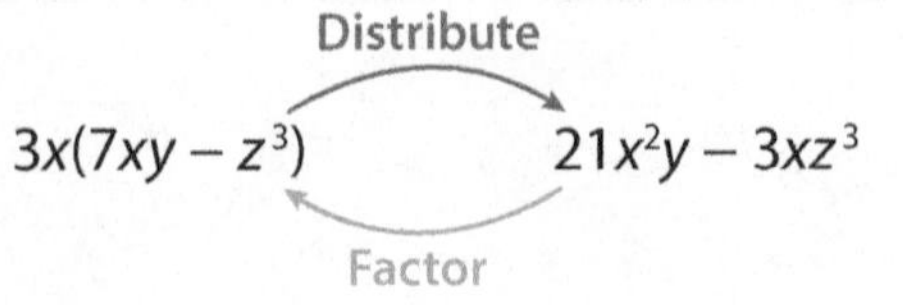

Figure 2.2. Distribution and Factoring

Factoring is the reverse of distributing: the first step is always to remove ("undistribute") the GCF of all the terms, if there is a GCF (besides 1). The GCF is the product of any constants and/or variables that <u>every</u> term shares. (For example, the GCF of $12x^3$, $15x^2$ and $6xy^2$ is $3x$ because $3x$ evenly divides all three terms.) This shared factor can be taken out of each term and moved to the outside of the parentheses, leaving behind a polynomial where each term is the original term divided by the GCF. (The remaining terms for the terms in the example would be $4x^2$, $5x$, and $2xy$.) It may be possible to factor the polynomial in the parentheses further, depending on the problem.

SAMPLE QUESTION

3) Expand the following expression: $5x(x^2 - 2c + 10)$

Answer:

$5x(x^2 - 2c + 10)$	
$(5x)(x^2) = 5x^3$	Distribute and multiply the term
$(5x)(-2c) = -10xc$	outside the parentheses to all three
$(5x)(10) = 50x$	terms inside the parentheses.
$= 5x^3 - 10xc + 50x$	

4)   **Expand the following expression:** $(x^2 - 5)(2x - x^3)$

Answer:

$(x^2 - 5)(2x - x^3)$	
$(x^2)(2x) = 2x^3$	
$(x^2)(-x^3) = -x^5$	Apply FOIL: first, outside, inside, and
$(-5)(2x) = -10x$	last.
$(-5)(-x^3) = 5x^3$	
$= 2x^3 - x^5 - 10x + 5x^3$	Combine like terms and put them in order.
$= -x^5 + 7x^3 - 10x$	

5)   **Factor the expression** $16z^2 + 48z$

Answer:

$16z^2 + 48z$	Both terms have a z, and 16 is a common factor of both 16 and 48. So the greatest common factor is $16z$. Factor out the GCF.
$= 16z(z + 3)$	

6)   **Factor the expression** $6m^3 + 12m^3n - 9m^2$

Answer:

$6m^3 + 12m^3n - 9m^2$	All the terms share the factor m^2, and 3 is the greatest common factor of 6, 12, and 9. So, the GCF is $3m^2$.
$= 3m^2(2m + 4mn - 3)$	

FACTORING TRINOMIALS

If the leading coefficient is $a = 1$, the trinomial is in the form $x^2 + bx + c$ and can often be rewritten in the factored form, as a product of two binomials: $(x + m)(x + n)$. Recall that the product of two binomials can be written in expanded form $x^2 + mx + nx + mn$. Equating this expression with $x^2 + bx + c$, the constant term c would have to equal the product mn. Thus, to work backward from the trinomial to the factored form, consider all the numbers m and n that multiply to make c. For example, to factor $x^2 + 8x + 12$, consider all the pairs that multiply to be 12 ($12 = 1 \times 12$ or 2×6 or 3×4). Choose the pair that will make the coefficient of the middle term (8) when added. In this example 2 and 6 add to 8, so making $m = 2$ and $n = 6$ in the expanded form gives:

$x^2 + 8x + 12 = x^2 + 2x + 6x + 12$	
$= (x^2 + 2x) + (6x + 12)$	Group the first two terms and the last two terms.
$= x(x + 6) + 2(x + 6)$	Factor the GCF out of each set of parentheses.
$= (x + 6)(x + 2)$	The two terms now have the common factor $(x + 6)$, which can be removed, leaving $(x + 2)$ and the original polynomial is factored.

In general:

$x^2 + bx + c = x^2 + mx + nx + mn$, where $c = mn$ and $b = m + n$	
$= (x^2 + mx) + (nx + mn)$	Group.
$= x(x + m) + n(x + m)$	Factor each group.
$= (x + m)(x + n)$	Factor out the common binomial.

Note that if none of the factors of c add to the value b, then the trinomial cannot be factored, and is called **prime**.

If the leading coefficient is not 1 ($a \neq 1$), first make sure that any common factors among the three terms are factored out. If the a-value is negative, factor out –1 first as well. If the a-value of the new polynomial in the parentheses is still not 1, follow this rule: Identify two values r and s that multiply to be ac and add to be b. Then write the polynomial in this form: $ax^2 + bx + c = ax^2 + rx + sx + c$, and proceed by grouping, factoring, and removing the common binomial as above.

There are a few special factoring cases worth memorizing: difference of squares, binomial squared, and the sum and difference of cubes.

- **Difference of squares** (each term is a square and they are subtracted):
  - $a^2 - b^2 = (a + b)(a - b)$
  - Note that a SUM of squares is never factorable.
- **Binomial squared**:
  - $a^2 + 2ab + b^2 = (a + b)(a + b) = (a + b)^2$
- **Sum and difference of cubes**:
  - $a^3 + b^3 = (a + b)(a^2 - ab + b^2)$
  - $a^3 - b^3 = (a - b)(a^2 + ab + b^2)$

▷ Note that the second factor in these factorizations will never be able to be factored further.

SAMPLE QUESTIONS

7)   Factor: $16x^2 + 52x + 30$

Answer:

$16x^2 + 52x + 30$	
$= 2(8x^2 + 26x + 15)$	Remove the GCF of 2.
$= 2(8x^2 + 6x + 20x + 15)$	To factor the polynomial in the parentheses, calculate $ac = (8)(15) = 120$, and consider all the pairs of numbers that multiply to be 120: 1×120, 2×60, 3×40, 4×30, 5×24, 6×20, 8×15, and 10×12. Of these pairs, choose the pair that adds to be the b-value 26 (6 and 20).
$= 2[(8x^2 + 6x) + (20x + 15)]$	Group.
$= 2[(2x(4x + 3) + 5(4x + 3)]$	Factor out the GCF of each group.
$= 2[(4x + 3)(2x + 5)]$	Factor out the common binomial.
$2(4x + 3)(2x + 5)$	

If there are no values r and s that multiply to be ac and add to be b, then the polynomial is prime and cannot be factored.

8)   Factor: $-21x^2 - x + 10$

Answer:

$-21x^2 - x + 10$	
$= -(21x^2 + x - 10)$	Factor out the negative.
$= -(21x^2 - 14x + 15x - 10)$	Factor the polynomial in the parentheses. $ac = 210$ and $b = 1$ The numbers 15 and −14 can be multiplied to get 210 and subtracted to get 1.
$= -[(21x^2 - 14x) + (15x - 10)]$	Group.
$= -[7x(3x - 2) + 5(3x - 2)]$	Factor out the GCF of each group.
$= -(3x - 2)(7x + 5)$	Factor out the common binomial.

LINEAR EQUATIONS

An **equation** states that two expressions are equal to each other. Polynomial equations are categorized by the highest power of the variables they contain: the highest power of any exponent of a linear equation is 1, a quadratic equation has a variable raised to the second power, a cubic equation has a variable raised to the third power, and so on.

SOLVING LINEAR EQUATIONS

Solving an equation means finding the value or values of the variable that make the equation true. To solve a linear equation, it is necessary to manipulate the terms so that the variable being solved for appears alone on one side of the equal sign while everything else in the equation is on the other side.

The way to solve linear equations is to "undo" all the operations that connect numbers to the variable of interest. Follow these steps:

1. Eliminate fractions by multiplying each side by the least common multiple of any denominators.

2. Distribute to eliminate parentheses, braces, and brackets.

3. Combine like terms.

4. Use addition or subtraction to collect all terms containing the variable of interest to one side, and all terms not containing the variable to the other side.

5. Use multiplication or division to remove coefficients from the variable of interest.

> **HELPFUL HINT**
>
> On multiple choice tests, it is often easier to plug the possible values into the equation and determine which solution makes the equation true than to solve the equation.

Sometimes there are no numeric values in the equation or there are a mix of numerous variables and constants. The goal is to solve the equation for one of the variables in terms of the other variables. In this case, the answer will be an expression involving numbers and letters instead of a numeric value.

SAMPLE QUESTIONS

9)   **Solve for *x*:** $\dfrac{100(x + 5)}{20} = 1$

Answer:

$$\frac{100(x + 5)}{20} = 1$$

$(20)\left(\frac{100(x+5)}{20}\right) = (1)(20)$ $100(x+5) = 20$	Multiply both sides by 20 to cancel out the denominator.
$100x + 500 = 20$	Distribute 100 through the parentheses.
$100x = -480$	"Undo" the +500 by subtracting 500 on both sides of the equation to isolate the variable term.
$x = \frac{-480}{100}$	"Undo" the multiplication by 100 by dividing by 100 on both sides to solve for x.
$x = \mathbf{-4.8}$	

10) **Solve for x: $2(x + 2)^2 - 2x^2 + 10 = 42$**

Answer:

$2(x + 2)^2 - 2x^2 + 10 = 42$	
$2(x + 2)(x + 2) - 2x^2 + 10 = 42$	Eliminate the exponents on the left side.
$2(x^2 + 4x + 4) - 2x^2 + 10 = 42$	Apply FOIL.
$2x^2 + 8x + 8 - 2x^2 + 10 = 42$	Distribute the 2.
$8x + 18 = 42$	Combine like terms on the left-hand side.
$8x = 24$	Isolate the variable. "Undo" +18 by subtracting 18 on both sides.
$\mathbf{x = 3}$	"Undo" multiplication by 8 by dividing both sides by 8.

11) **Solve the equation for D: $\frac{A(3B + 2D)}{2N} = 5M - 6$**

Answer:

$\frac{A(3B + 2D)}{2N} = 5M - 6$	
$3AB + 2AD = 10MN - 12N$	Multiply both sides by $2N$ to clear the fraction, and distribute the A through the parentheses.
$2AD = 10MN - 12N - 3AB$	Isolate the term with the D in it by moving $3AB$ to the other side of the equation.
$\mathbf{D = \frac{(10MN - 12N - 3AB)}{2A}}$	Divide both sides by $2A$ to get D alone on the right-hand side.

GRAPHS OF LINEAR EQUATIONS

The most common way to write a linear equation is **slope-intercept form**, $y = mx + b$. In this equation, m is the slope, which describes how steep the line is, and b is the y-intercept. Slope is often described as "rise over run" because it is calculated as the difference in y-values (rise) over the difference in x-values (run). The slope of the line is also the rate of change of the dependent variable y with respect to the independent variable x. The y-intercept is the point where the line crosses the y-axis, or where x equals zero.

To graph a linear equation, identify the y-intercept and place that point on the y-axis. If the slope is not written as a fraction, make it a fraction by writing it over $1\left(\frac{m}{1}\right)$. Then use the slope to count up (or down, if negative) the "rise" part of the slope and over the "run" part of the slope to find a second point. These points can then be connected to draw the line.

To find the equation of a line, identify the y-intercept, if possible, on the graph and use two easily identifiable points to find the slope. If the y-intercept is not easily identified, identify the slope by choosing easily identifiable points; then choose one point on the graph, plug the point and the slope values into the equation, and solve for the missing value b.

- standard form: $Ax + By = C$
- $m = -\frac{A}{B}$
- x-intercept $= \frac{C}{A}$
- y-intercept $= \frac{C}{B}$

Another way to express a linear equation is standard form: $Ax + By = C$. In order to graph equations in this form, it is often easiest to convert them to point-slope form. Alternately, it is easy to find the x- or y-intercept from this form, and once these two points are known, a line can be drawn through them. To find the x-intercept, simply make $y = 0$ and solve for x. Similarly, to find the y-intercept, make $x = 0$ and solve for y.

SAMPLE QUESTIONS

12) **What is the slope of the line whose equation is $6x - 2y - 8 = 0$?**

Answer:

$6x - 2y - 8 = 0$

$-2y = -6x + 8$ $y = \frac{-6x + 8}{-2}$ $y = 3x - 4$	Rearrange the equation into slope-intercept form by solving the equation for y.
$m = 3$	The slope is 3, the value attached to x.

13)   **What is the equation of the following line?**

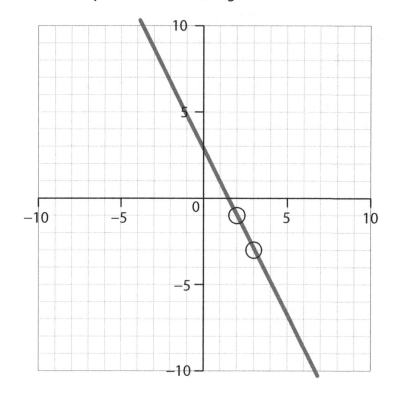

Answer:

$b = 3$	The y-intercept can be identified on the graph as $(0, 3)$.
$m = \frac{(-3) - (-1)}{3 - 2} = \frac{-2}{1} = -2$	To find the slope, choose any two points and plug the values into the slope equation. The two points chosen here are $(2, -1)$ and $(3, -3)$.
$y = -2x + 3$	Replace m with -2 and b with 3 in $y = mx + b$.

14)   **Write the equation of the line which passes through the points $(-2, 5)$ and $(-5, 3)$.**

Answer:

$(-2, 5)$ and $(-5, 3)$

$m = \dfrac{3-5}{(-5)-(-2)}$ $= \dfrac{-2}{-3}$ $= \dfrac{2}{3}$	Calculate the slope.
$5 = \dfrac{2}{3}(-2) + b$ $5 = \dfrac{-4}{3} + b$ $b = \dfrac{19}{3}$	To find b, plug into the equation $y = mx + b$ the slope for m and a set of points for x and y.
$y = \dfrac{2}{3}x + \dfrac{19}{3}$	Replace m and b to find the equation of the line.

15) **What is the equation of the following graph?**

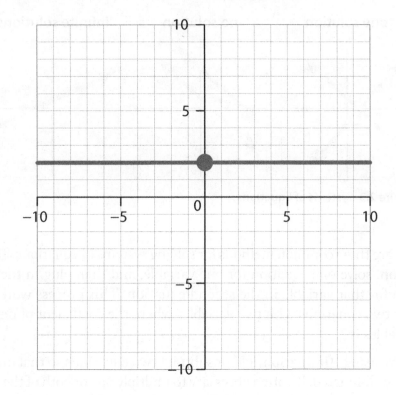

Answer:

$y = 0x + 2$, **or** $y = 2$

The line has a rise of 0 and a run of 1, so the slope is $\dfrac{0}{1} = 0$. There is no x-intercept. The y-intercept is $(0, 2)$, meaning that the b-value in the slope-intercept form is 2.

SYSTEMS OF LINEAR EQUATIONS

Systems of equations are sets of equations that include two or more variables. These systems can only be solved when there are at least as many equations as there are variables. Systems involve working with more than one equation to solve for more than one variable. For a system of linear equations, the solution to the system is the set of values for the variables that satisfies every equation in the system. Graphically, this will be the point where every line meets. If the lines are parallel (and hence do not intersect), the system will have no solution. If the lines are multiples of each other, meaning they share all coordinates, then the system has infinitely many solutions (because every point on the line is a solution).

> **HELPFUL HINT**
>
> Plug answers back into both equations to ensure the system has been solved properly.

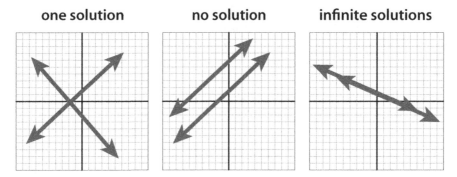

one solution   **no solution**   **infinite solutions**

Figure 2.3. Systems of Equations

There are three common methods for solving systems of equations. To perform **substitution**, solve one equation for one variable, and then plug in the resulting expression for that variable in the second equation. This process works best for systems of two equations with two variables where the coefficient of one or more of the variables is 1.

To solve using **elimination**, add or subtract two equations so that one or more variables are eliminated. It's often necessary to multiply one or both of the equations by a scalar (constant) in order to make the variables cancel. Equations can be added or subtracted as many times as necessary to find each variable.

Yet another way to solve a system of linear equations is to use a **matrix equation**. In the matrix equation $AX = B$, A contains the system's coefficients, X contains the variables, and B contains the constants (as shown below). The matrix equation can then be solved by multiplying B by the inverse of A: $\boldsymbol{X = A^{-1}B}$

$$\begin{matrix} ax + by = e \\ cx + dy = f \end{matrix} \rightarrow A = \begin{bmatrix} a & b \\ c & d \end{bmatrix} \quad X = \begin{bmatrix} x \\ y \end{bmatrix} \quad B = \begin{bmatrix} e \\ f \end{bmatrix} \rightarrow AX = B$$

This method can be extended to equations with three or more variables. Technology (such as a graphing calculator) is often employed when solving using this method if more than two variables are involved.

SAMPLE QUESTIONS

16) **Solve for x and y:**

$2x - 4y = 28$

$4x - 12y = 36$

Answer:

$2x - 4y = 28$ $x = 2y + 14$	Solve the system with substitution. Solve one equation for one variable.
$4x - 12y = 36$ $4(2y + 14) - 12y = 36$ $8y + 56 - 12y = 36$ $-4y = -20$ $y = 5$	Plug in the resulting expression for x in the second equation and simplify.
$2x - 4y = 28$ $2x - 4(5) = 28$ $2x - 20 = 28$ $2x = 48$ $x = 24$ The answer is $y = 5$ and $x = 24$ or **(24, 5)**.	Plug the solved variable into either equation to find the second variable.

17) **Solve for the system for x and y:**

$3 = -4x + y$

$16x = 4y + 2$

Answer:

$3 = -4x + y$ $y = 4x + 3$	Isolate the variable in one equation.
$16x = 4y + 2$ $16x = 4(4x + 3) + 2$ $16x = 16x + 12 + 2$ $0 = 14$ **No solution exists.**	Plug the expression into the second equation. Both equations have slope 4. This means the graphs of the equations are parallel lines, so no intersection (solution) exists.

18) Solve the system of equations:

$$6x + 10y = 18$$

$$4x + 15y = 37$$

Answer:

Because solving for *x* or *y* in either equation will result in messy fractions, this problem is best solved using elimination. The goal is to eliminate one of the variables by making the coefficients in front of one set of variables the same, but with different signs, and then adding both equations.

$6x + 10y = 18 \xrightarrow[(-2)]{} -12x - 20y = -36$ $4x + 15y = 37 \xrightarrow[(3)]{} 12x + 45y = 111$	To eliminate the *x*'s in this problem, find the least common multiple of coefficients 6 and 4. The smallest number that both 6 and 4 divide into evenly is 12. Multiply the top equation by −2, and the bottom equation by 3.
$25y = 75$	Add the two equations to eliminate the *x*'s.
$y = 3$	Solve for *y*.
$6x + 10(3) = 18$ $6x + 30 = 18$ $x = -2$	Replace *y* with 3 in either of the original equations.
The solution is **(−2, 3)**.	

19) Solve the following systems of equations using matrix arithmetic:

$$2x - 3y = -5$$

$$3x - 4y = -8$$

Answer:

$\begin{bmatrix} 2 & -3 \\ 3 & -4 \end{bmatrix} \begin{bmatrix} x \\ y \end{bmatrix} = \begin{bmatrix} -5 \\ -8 \end{bmatrix}$	Write the system in matrix form, ***AX = B***.
$\begin{bmatrix} 2 & -3 \\ 3 & -4 \end{bmatrix}^{-1}$ $= \dfrac{1}{(2)(-4) - (-3)(3)} \begin{bmatrix} -4 & 3 \\ -3 & 2 \end{bmatrix} = \begin{bmatrix} -4 & 3 \\ -3 & 2 \end{bmatrix}$	Calculate the inverse of Matrix ***A***.
$\begin{bmatrix} x \\ y \end{bmatrix} = \begin{bmatrix} -4 & 3 \\ -3 & 2 \end{bmatrix} \begin{bmatrix} -5 \\ -8 \end{bmatrix} = \begin{bmatrix} -4 \\ -1 \end{bmatrix}$	Multiply ***B*** by the inverse of ***A***.
x = −4 ***y = −1***	Match up the 2 × 1 matrices to identify *x* and *y*.

BUILDING EQUATIONS

In word problems, it is often necessary to translate a verbal description of a relationship into a mathematical equation. No matter the problem, this process can be done using the same steps:

1. Read the problem carefully and identify what value needs to be solved for.

2. Identify the known and unknown quantities in the problem, and assign the unknown quantities a variable.

3. Create equations using the variables and known quantities.

4. Solve the equations.

5. Check the solution: Does it answer the question asked in the problem? Does it make sense?

> **HELPFUL HINT**
>
> Use the acronym STAR to remember word-problem strategies: Search the problem, Translate into an expression or equation, Answer, and Review.

SAMPLE QUESTIONS

20) A school is holding a raffle to raise money. There is a $3 entry fee, and each ticket costs $5. If a student paid $28, how many tickets did he buy?

Answer:

Number of tickets = x Cost per ticket = 5 Cost for x tickets = $5x$ Total cost = 28 Entry fee = 3	Identify the quantities.
$5x + 3 = 28$	Set up equations. The total cost for x tickets will be equal to the cost for x tickets plus the $3 flat fee.
$5x + 3 = 28$ $5x = 25$ $x = 5$	Solve the equation for x.

The student bought **5 tickets**.

21) Kelly is selling shirts for her school swim team. There are two prices: a student price and a nonstudent price. During the first week of the sale, Kelly raised $84 by selling 10 shirts to students and 4 shirts to nonstudents. She earned $185 in the second week by selling 20 shirts to students and 10 shirts to nonstudents. What is the student price for a shirt?

Answer:

Student price = s Nonstudent price = n	Assign variables.
$10s + 4n = 84$ $20s + 10n = 185$	Create two equations using the number of shirts Kelly sold and the money she earned.
$10s + 4n = 84$ $10n = -20s + 185$ $n = -2s + 18.5$ $10s + 4(-2s + 18.5) = 84$ $10s - 8s + 74 = 84$ $2s + 74 = 84$ $2s = 10$ $s = 5$	Solve the system of equations using substitution.
The student cost for shirts is **$5**.	

LINEAR INEQUALITIES

An inequality shows the relationship between two expressions, much like an equation. However, the equal sign is replaced with an inequality symbol that expresses the following relationships:

- ▶ < less than
- ▶ ≤ less than or equal to
- ▶ > greater than
- ▶ ≥ greater than or equal to

Inequalities are read from left to right. For example, the inequality $x \leq 8$ would be read as "x is less than or equal to 8," meaning x has a value smaller than or equal to 8. The set of solutions of an inequality can be expressed using a number line. The shaded region on the number line represents the set of all the numbers that

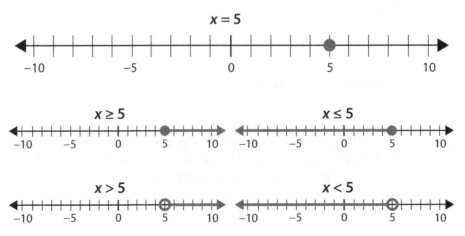

Figure 2.4. Inequalities on a Number Line

make an inequality true. One major difference between equations and inequalities is that equations generally have a finite number of solutions, while inequalities generally have infinitely many solutions (an entire interval on the number line containing infinitely many values).

Linear inequalities can be solved in the same way as linear equations, with one exception. When multiplying or dividing both sides of an inequality by a negative number, the direction of the inequality sign must reverse—"greater than" becomes "less than" and "less than" becomes "greater than."

SAMPLE QUESTIONS

22) **Solve for z: $3z + 10 < -z$**

Answer:

$3z + 10 < -z$	
$3z < -z - 10$	Collect nonvariable terms to one side.
$4z < -10$	Collect variable terms to the other side.
$z < -2.5$	Isolate the variable.

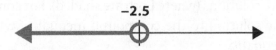

23) **Solve for x: $2x - 3 > 5(x - 4) - (x - 4)$**

Answer:

$2x - 3 > 5(x - 4) - (x - 4)$	
$2x - 3 > 5x - 20 - x + 4$	Distribute 5 through the parentheses and −1 through the parentheses.
$2x - 3 > 4x - 16$	Combine like terms.
$-2x > -13$	Collect x-terms to one side, and constant terms to the other side.
$x < 6.5$	Divide both sides by −2; since dividing by a negative, reverse the direction of the inequality.

COMPOUND INEQUALITIES

Compound inequalities have more than one inequality expression. Solutions of compound inequalities are the sets of all numbers that make *all* the inequalities true. Some compound inequalities may not have any solutions, some will have solutions that contain some part of the number line, and some will have solutions that include the entire number line.

Table 2.1. Unions and Intersections

Inequality	Meaning in Words	Number Line
$a < x < b$	All values x that are greater than a and less than b	
$a \leq x \leq b$	All values x that are greater than or equal to a and less than or equal to b	
$x < a \text{ or } x > b$	All values of x that are less than a or greater than b	
$x \leq a \text{ or } x \geq b$	All values of x that are less than or equal to a or greater than or equal to b	

Compound inequalities can be written, solved, and graphed as two separate inequalities. For compound inequalities in which the word *and* is used, the solution to the compound inequality will be the set of numbers on the number line where both inequalities have solutions (where both are shaded). For compound inequalities where *or* is used, the solution to the compound inequality will be *all* the shaded regions for *either* inequality.

SAMPLE QUESTIONS

24) Solve the compound inequalities: $2x + 4 < -18 \text{ or } 4(x + 2) > 18$

Answer:

$2x + 4 < -10 \text{ or } 4(x + 2) > 18$	
$2x < -14 \qquad 4x + 8 > 18$	
$x < -7 \qquad\quad 4x > 10$	Solve each inequality independently.
$\qquad\qquad\quad x > 2.5$	

The solution to the original compound inequality is **the set of all x for which x < −7 or x > 2.5.**

25) Solve the inequality: $-1 \leq 3(x + 2) - 1 \leq x + 3$

Answer:

$-1 \leq 3(x + 2) - 1 \leq x + 3$

$-1 \le 3(x + 2) - 1$ *and* $3(x + 2) - 1 \le x + 3$	Break up the compound inequality into two inequalities.
$-1 \le 3x + 6 - 1$ $3x + 6 - 1 \le x + 3$ $-6 \le 3x$ $2x \le -2$ $-2 \le x$ and $x \le -1$	Solve separately.
$\mathbf{-2 \le x \le -1}$	The only values of x that satisfy *both* inequalities are the values between −2 and −1 (inclusive).

GRAPHING LINEAR INEQUALITIES IN TWO VARIABLES

Linear inequalities in two variables can be graphed in much the same way as linear equations. Start by graphing the corresponding equation of a line (temporarily replace the inequality with an equal sign, and then graph). This line creates a boundary line of two half-planes. If the inequality is a "greater/less than," the boundary should not be included and a dotted line is used. A solid line is used to indicate that the boundary should be included in the solution when the inequality is "greater/less than or equal to."

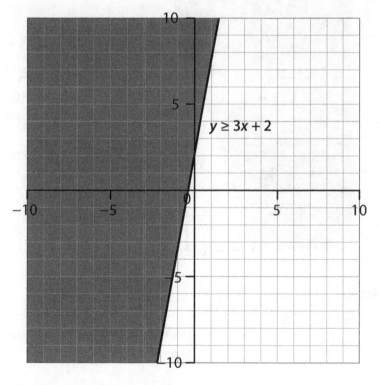

Figure 2.5. Graphing Inequalities

One side of the boundary is the set of all points (x,y) that make the inequality true. This side is shaded to indicate that all these values are solutions. If y is greater than the expression containing *x*, shade above the line; if it is less than, shade below. A point can also be used to check which side of the line to shade.

A set of two or more linear inequalities is a **system of inequalities**. Solutions to the system are all the values of the variables that make every inequality in the system true. Systems of inequalities are solved graphically by graphing all the inequalities in the same plane. The region where all the shaded solutions overlap is the solution to the system.

SAMPLE QUESTIONS

26)   **What is the inequality represented on the graph below?**

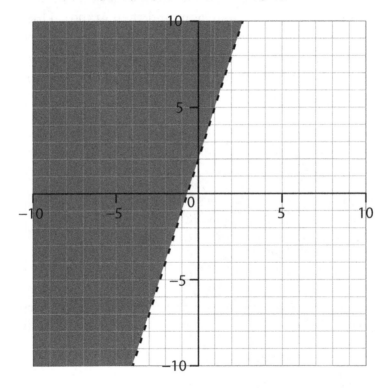

Answer:

y-intercept: (0,2)

slope: 3

$y = 3x + 2$

Determine the equation of the boundary line.

Replace the equal sign with the appropriate inequality: the line is dotted and the shading is above the line, indicating that the symbol should be "greater than." Check a point: for example (1, 5) is a solution since 5 > 3(−1) + 2.

$$y > 3x + 2$$

27) **Graph the following inequality: $3x + 6y \leq 12$.**

  Answer:

$3x + 6y \leq 12$ $3(0) + 6y = 12$ $y = 2$ y-intercept: $(0, 2)$ $3x + 6(0) \leq 12$ $x = 4$ x-intercept: $(4, 0)$	Find the x- and y-intercepts.

  Graph the line using the intercepts, and shade below the line.

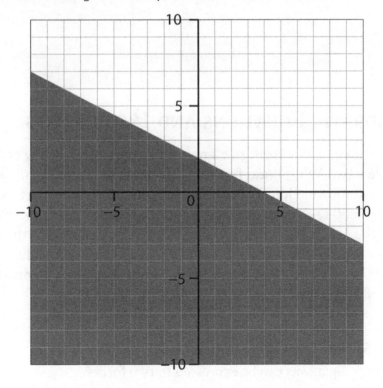

28) **Graph the system of inequalities: $-x + y \leq 1, x \geq -1, y > 2x - 4$**

Answer:

To solve the system, graph all three inequalities in the same plane; then identify the area where the three solutions overlap. All points (x, y) in this area will be solutions to the system since they satisfy all three inequalities.

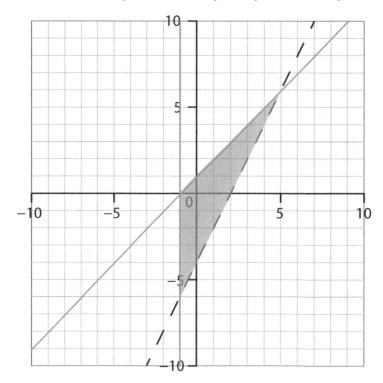

QUADRATIC EQUATIONS AND INEQUALITIES

Quadratic equations are degree 2 polynomials; the highest power on the dependent variable is two. While linear functions are represented graphically as lines, the graph of a quadratic function is a **parabola**. The graph of a parabola has three important components. The **vertex** is where the graph changes direction. In the parent graph $y = x^2$, the origin $(0,0)$ is the vertex. The **axis of symmetry** is the vertical line that cuts the graph into two equal halves. The line of symmetry always passes through the vertex. On the parent graph, the y-axis is the axis of symmetry. The **zeros** or **roots** of the quadratic are the x-intercepts of the graph.

FORMS OF QUADRATIC EQUATIONS

Quadratic equations can be expressed in two forms:

▸ **Standard form:** $y = ax^2 + bx + c$

Axis of symmetry: $x = -\frac{b}{2a}$ Vertex: $(-\frac{b}{2a}, f(-\frac{b}{2a}))$

▸ **Vertex form:** $y = a(x - h)^2 + k$

Vertex: (h,k) Axis of symmetry: $x = h$

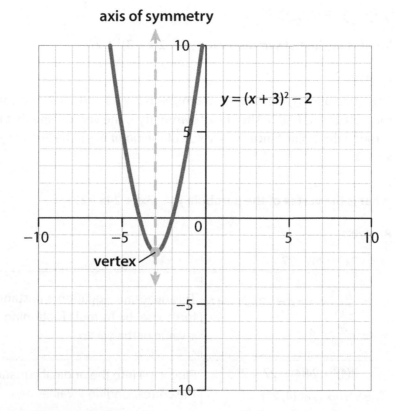

axis of symmetry

$y = (x + 3)^2 - 2$

vertex

Figure 2.6. Parabola

In both equations, the sign of *a* determines which direction the parabola opens: if *a* is positive, then it opens upward; if *a* is negative, then it opens downward. The wideness or narrowness is also determined by *a*. If the absolute value of *a* is less than one (a proper fraction), then the parabola will get wider the closer ｜*a*｜ is to zero. If the absolute value of *a* is greater than one, then the larger ｜*a*｜ becomes, the narrower the parabola will be.

Equations in vertex form can be converted to standard form by squaring out the $(x - h)^2$ part (using FOIL), distributing the *a*, adding *k*, and simplifying the result.

Equations can be converted from standard form to vertex form by **completing the square.** Take an equation in standard form, $y = ax^2 + bc + c$.

1. Move *c* to the left side of the equation.
2. Divide the entire equation through by *a* (to make the coefficient of x^2 be 1).
3. Take half of the coefficient of *x*, square that number, and then add the result to both sides of the equation.
4. Convert the right side of the equation to a perfect binomial squared, $(x + m)^2$.
5. Isolate *y* to put the equation in proper vertex form.

SAMPLE QUESTIONS

29) **What is the line of symmetry for $y = -2(x + 3)^2 + 2$?**

Answer:

This quadratic is given in vertex form, with $h = -3$ and $k = 2$. The vertex of this equation is $(-3, 2)$. The line of symmetry is the vertical line that passes through this point. Since the x-value of the point is -3, the line of symmetry is **$x = -3$**.

30) **What is the vertex of the parabola $y = -3x^2 + 24x - 27$?**

Answer:

$y = -3x^2 + 24x - 27$	
$x = -\dfrac{b}{2a}$ where $a = -3$, $b = 24$ $x = -\dfrac{24}{2(-3)} = 4$	This quadratic equation is in standard form. Use the formula for finding the x-value of the vertex.
$y = -3(4)^2 + 24(4) - 27 = 21$ The vertex is at **$(4, 21)$**.	Plug $x = 4$ into the original equation to find the corresponding y-value.

31) **Write $y = -3x^2 + 24x - 27$ in vertex form by completing the square.**

Answer:

$y = -3x^2 + 24x - 27$	
$y + 27 = -3x^2 + 24x$	Move c to the other side of the equation.
$\dfrac{y}{-3} - 9 = x^2 - 8x$	Divide through by a (-3 in this example).
$\dfrac{y}{-3} - 9 + 16 = x^2 - 8x + 16$	Take half of the new b, square it, and add that quantity to both sides: $\frac{1}{2}(-8) = -4$. Squaring it gives $(-4)^2 = 16$.
$\dfrac{y}{-3} + 7 = (x - 4)^2$	Simplify the left side, and write the right side as a binomial squared.
$y = -3(x - 4)^2 + 21$	Subtract 7, and then multiply through by -3 to isolate y.

Solving Quadratic Equations

Solving the quadratic equation $ax^2 + bx + c = 0$ finds x-intercepts of the parabola (by making $y = 0$). These are also called the **roots** or **zeros** of the quadratic function. A quadratic equation may have zero, one, or two real solutions. There are several ways of finding the zeros. One way is to factor the quadratic into a product of two binomials, and then use the zero product property. (If $m \times n = 0$, then either $m = 0$ or $n = 0$.) Another way is to complete the square and square root both sides. One way that works every time is to memorize and use the **quadratic formula**:

$$x = \frac{-b \pm \sqrt{b^2 - 4ac}}{2a}$$

The a, b, and c come from the standard form of quadratic equations above. (Note that to use the quadratic equation, the right-hand side of the equation must be equal to zero.)

The part of the formula under the square root radical ($b^2 - 4ac$) is known as the **discriminant**. The discriminant tells how many and what type of roots will result without actually calculating the roots.

> **HELPFUL HINT**
>
> With all graphing problems, putting the function into the y = window of a graphing calculator will aid the process of elimination when graphs are examined and compared to answer choices with a focus on properties like axis of symmetry, vertices, and roots of formulas.

Table 2.2. Discriminants

If $b^2 - 4ac$ is	there will be	and the parabola
zero	only 1 real root	has its vertex on the x-axis
positive	2 real roots	has **two** x-intercepts
negative	0 real roots 2 complex roots	has **no** x-intercepts

SAMPLE QUESTIONS

32) **Find the zeros of the quadratic equation: $y = -(x + 3)^2 + 1$.**

Answer:

Method 1: Make $y = 0$; isolate x by square rooting both sides:

$0 = -(x + 3)^2 + 1$	Make $y = 0$.
$-1 = -(x + 3)^2$	Subtract 1 from both sides.
$1 = (x + 3)^2$	Divide by -1 on both sides.
$(x + 3) = \pm 1$	Square root both sides. Don't forget to write plus OR minus 1.
$(x + 3) = 1$ or $(x + 3) = -1$	Write two equations using $+1$ and -1.
$x = -2$ or $x = -4$	Solve both equations. These are the zeros.

Method 2: Convert vertex form to standard form, and then use the quadratic formula.

$y = -(x + 3)^2 + 1$ $y = -(x^2 + 6x + 9) + 1$ $y = -x^2 - 6x - 8$	Put the equation in standard form by distributing and combining like terms.
$x = \dfrac{-b \pm \sqrt{(b^2 - 4ac)}}{2a}$ $x = \dfrac{-(-6) \pm \sqrt{(-6)^2 - 4(-1)(-8)}}{2(-1)}$ $x = \dfrac{6 \pm \sqrt{36 - 32}}{-2}$ $x = \dfrac{6 \pm \sqrt{4}}{-2}$ $x = -4, -2$	Find the zeros using the quadratic formula.

33) **Find the root(s) for:** $z^2 - 4z + 4 = 0$

Answer:

This polynomial can be factored in the form $(z - 2)(z - 2) = 0$, so the only root is $z = 2$. There is only one x-intercept, and the vertex of the graph is *on* the x-axis.

34) **Write a quadratic function that has zeros at** $x = -3$ **and** $x = 2$ **that passes through the point** $(-2, 8)$.

Answer:

If the quadratic has zeros at $x = -3$ and $x = 2$, then it has factors of $(x + 3)$ and $(x - 2)$. The quadratic function can be written in the factored form $y = a(x + 3)(x - 2)$. To find the a-value, plug in the point $(-2, 8)$ for x and y:

$8 = a(-2 + 3)(-2 - 2)$

$8 = a(-4)$

$a = -2$

The quadratic function is $y = -2(x + 3)(x - 2)$.

GRAPHING QUADRATIC EQUATIONS

The final expected quadratic skills are graphing a quadratic function given its equation and determining the equation of a quadratic function from its graph. The equation's form determines which quantities are easiest to obtain:

Table 2.3 Obtaining Quantities from Quadratic Functions

Name of Form	Equation of Quadratic	Easiest Quantity to Find	How to Find Other Quantities
vertex form	$y = a(x - h)^2 + k$	vertex at (h, k) and axis of symmetry $x = h$	Find zeros by making $y = 0$ and solving for x.
factored form	$y = a(x - m)(x - n)$	x – intercepts at $x = m$ and $x = n$	Find axis of symmetry by averaging m and n: $x = \frac{m + n}{2}$. This is also the x-value of the vertex.
standard form	$y = ax^2 + bx + c$	y – intercept at $(0, c)$	Find axis of symmetry and x-value of the vertex using $x = \frac{-b}{2a}$. Find zeros using quadratic formula.

To graph a quadratic function, first determine if the graph opens up or down by examining the a-value. Then determine the quantity that is easiest to find based on the form given, and find the vertex. Then other values can be found, if necessary, by choosing x-values and finding the corresponding y-values. Using symmetry instantly doubles the number of points that are known.

Given the graph of a parabola, the easiest way to write a quadratic equation is to identify the vertex and insert the h- and k-values into the vertex form of the equation. The a-value can be determined by finding another point the graph goes through, plugging these values in for x and y, and solving for a.

SAMPLE QUESTIONS

35) **Graph the quadratic $y = 2(x - 3)^2 + 4$.**

Answer:

Start by marking the vertex at $(3, 4)$ and recognizing this parabola opens upward. The line of symmetry is $x = 3$. Now, plug in an easy value for x to get one point on the curve; then use symmetry to find another point. In this case, choose $x = 2$ (one unit to the left of the line of symmetry) and solve for y:

$$y = 2(2 - 3)^2 + 4$$

$$y = 2(1) + 4$$

$$y = 6$$

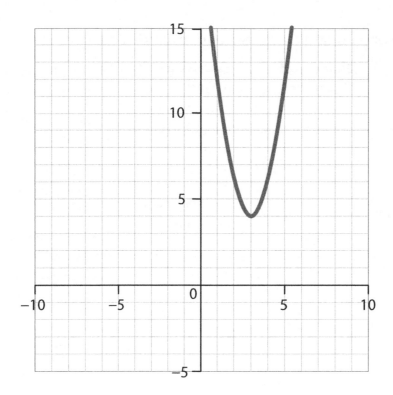

Thus the point $(2, 6)$ is on the curve. Then use symmetry to find the corresponding point one unit to the right of the line of symmetry, which must also have a y value of 6. This point is $(4, 6)$. Draw a parabola through the points.

36)   **What is the vertex form of the equation shown on the following graph?**

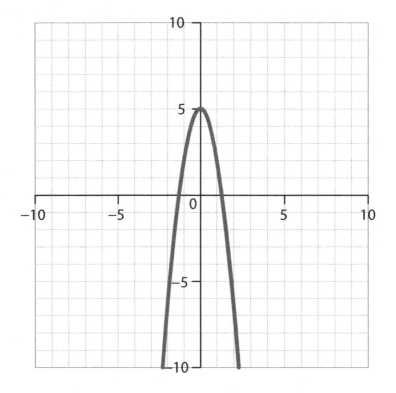

Answer:

$(h,k) = (0,5)$ $y = a(x - h)^2 + k$ $y = a(x - 0)^2 + 5$ $y = ax^2 + 5$	Locate the vertex and plug values for h and k into the vertex form of the quadratic equation.
$(x,y) = (1,2)$ $y = ax^2 + 5$ $2 = a(1)^2 + 5$ $a = -3$	Choose another point on the graph to plug into this equation to solve for a.
$\mathbf{y = -3x^2 + 5}$	Plug a into the vertex form of the equation.

QUADRATIC INEQUALITIES

Quadratic inequalities with two variables, such as $y < (x + 3)^2 - 2$ can be graphed much like linear inequalities: graph the equation by treating the inequality symbol as an equal sign, then shade the graph. Shade above the graph when y is greater is than, and below the graph when y is less than.

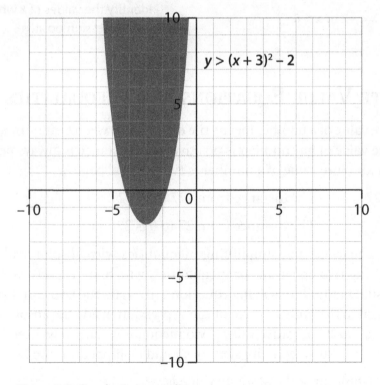

Figure 2.7. Quadratic Inequality

Quadratic inequalities with only one variable, such as $x^2 - 4x > 12$, can be solved by first manipulating the inequality so that one side is zero. The zeros can then be found and used to determine where the inequality is greater than zero (positive) or less than zero (negative). Often it helps to set up intervals on a number line and test a value within each range created by the zeros to identify the values that create positive or negative values.

SAMPLE QUESTION

37) **Find the values of x such that $x^2 - 4x > 12$.**

Answer:

$x^2 - 4x = 12$ $x^2 - 4x - 12 = 0$ $(x + 2)(x - 6) = 0$ $x = -2, 6$		Find the zeros of the inequality.

x	$(x + 2)(x - 6)$	
$-\infty < x < -2$	$+$	Create a table or number line with the intervals created by the zeros. Use a test value to determine whether the expression is positive or negative.
$-2 < x < 6$	$-$	
$6 < x < \infty$	$+$	

$x < -2$ or $x > 6$	Identify the values of x which make the expression positive.

ABSOLUTE VALUE EQUATIONS AND INEQUALITIES

The **absolute value** of a number means the distance between that number and zero. The absolute value of any number is positive since distance is always positive. The notation for absolute value of a number is two vertical bars:

$|-27| = 27$ The distance from –27 to 0 is 27.

$|27| = 27$ The distance from 27 to 0 is 27.

Solving equations and simplifying inequalities with absolute values usually requires writing two equations or inequalities, which are then solved separately using the usual methods of solving equations. To write the two equations, set one equation equal to the positive value of the expression inside the absolute value and the other equal to the negative value. Two inequalities can be written in the same manner. However, the inequality symbol should be flipped for the negative value.

The formal definition of the absolute value is

$$|x| = \begin{cases} -x, & x < 0 \\ x, & x \geq 0 \end{cases}$$

This is true because whenever x is negative, the opposite of x is the answer (for example, $|-5| = -(-5) = 5$, but when x is positive, the answer is just x. This type of function is called a **piece-wise function**. It is defined in two (or more) distinct pieces.

To graph the absolute value function, graph each piece separately. When $x < 0$ (that is, when it is negative), graph the line $y = -x$. When $x > 0$ (that is, when x is positive), graph the line $y = x$. This creates a V-shaped graph that is the parent function for absolute value functions.

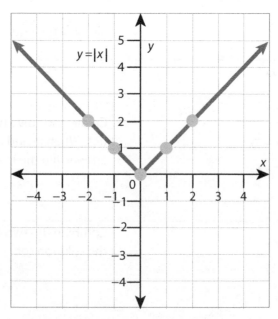

Figure 2.8. Absolute Value Parent Function

SAMPLE QUESTIONS

38) **Solve for x: |x − 3| = 27**

Answer:

Set the quantity inside the parentheses equal to 27 or −27, and solve:

$x - 3 = 27$ | $x - 3 = -27$

$x = 30$ | **$x = -24$**

39) **Solve for r: $\frac{|r - 7|}{5} = 27$**

Answer:

The first step is to isolate the absolute value part of the equation. Multiplying both sides by 5 gives:

$|r - 7| = 135$

If the quantity in the absolute value bars is 135 or −135, then the absolute value would be 135:

$r - 7 = 135$ | $r - 7 = -135$

$r = 142$ | **$r = -128$**

40) Find the solution set for the following inequality: $\left|\frac{3x}{7}\right| \geq 4 - x$.

Answer:

$\left|\frac{3x}{7}\right| \geq 4 - x$	
$\frac{|3x|}{7} \geq 4 - x$ $\|3x\| \geq 28 - 7x$	Simplify the equation.
$3x \geq 28 - 7x$ $10x \geq 28$ $x \geq \frac{28}{10}$ $-(3x) \leq 28 - 7x$ $-3x \leq 28 - 7x$ $4x \leq 28$ $x \leq 7$	Create and solve two inequalities. When including the negative answer, flip the inequality.
$\frac{28}{10} \leq x \leq 7$	Combine the two answers to find the solution set.

FUNCTIONS

WORKING WITH FUNCTIONS

Functions can be thought of as a process: when something is put in, an action (or operation) is performed, and something different comes out. A **function** is a relationship between two quantities (for example x and y) in which, for every value of the independent variable (usually x), there is exactly one value of the dependent variable (usually y). Briefly, each input has *exactly one* output. Graphically this means the graph passes the **vertical line test**: anywhere a vertical line is drawn on the graph, the line hits the curve at exactly one point.

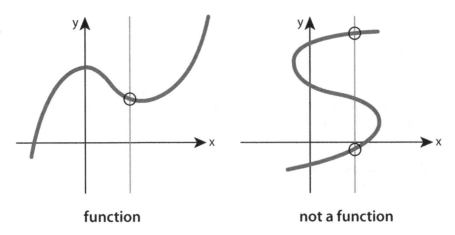

function   not a function

Figure 2.9. Vertical Line Test

The notation $f(x)$ or $g(t)$, etc., is often used when a function is being considered. This is **function notation.** The input value is x and the output value y is written as $y = f(x)$. Thus, $f(2)$ represents the output value (or y value) when $x = 2$, and $f(2) = 5$ means that when $x = 2$ is plugged into the $f(x)$ function, the output (y value) is 5. In other words, $f(2) = 5$ represents the point $(2,5)$ on the graph of $f(x)$.

Every function has an **input domain** and **output range**. The domain is the set of all the possible x values that can be used as input values (these are found along the horizontal axis on the graph), and the range includes all the y values or output values that result from applying $f(x)$ (these are found along the vertical axis on the graph). Domain and range are usually intervals of numbers and are often expressed as inequalities, such as $x < 2$ (the domain is all values less than 2) or $3 < x < 15$ (all values between 3 and 15).

Interval notation can also be used to show domain and range. Round brackets indicate that an end value is not included, and square brackets show that it is. The symbol ∪ means *or*, and the symbol ∩ means *and*. For example, the statement (–infinity,4) ∪ (4,infinity) describes the set of all real numbers except 4.

A function $f(x)$ is **even** if $f(-x) = f(x)$. Even functions have symmetry across the y-axis. An example of an even function is the parent quadratic $y = x^2$, because any value of x (for example, 3) and its opposite $-x$ (for example, –3) have the same y value (for example, $3^2 = 9$ and $(-3)^2 = 9$). A function is **odd** if $f(-x) = -f(x)$. Odd functions have symmetry about the origin. For example, $f(x) = x^3$ is an odd function because $f(3) = 27$, and $f(-3) = -27$. A function may be even, odd, or neither.

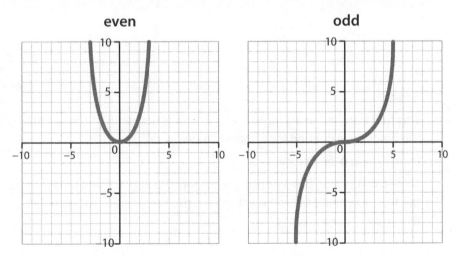

Figure 2.10. Even and Odd Functions

SAMPLE QUESTIONS

41) **Evaluate: $f(4)$ if $f(x) = x^3 - 2x + \sqrt{x}$**

Answer:

$f(x) = x^3 - 2x + \sqrt{x}$	
$f(4) = (4)^3 - 2(4) + \sqrt{(4)}$	Plug in 4.
$= 64 - 8 + 2$ $= \mathbf{58}$	Follow the PEMDAS order of operations.

42) **What are the domain and range of the following function?**

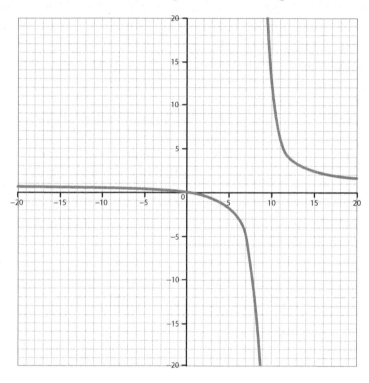

Answer:

This function has an asymptote at $x = 9$, so is not defined there. Otherwise, the function is defined for all other values of x.

D: $-\infty < x < \mathbf{9}$ or $\mathbf{9} < x < \infty$

Since the function has a horizontal asymptote at $y = 1$ that it never crosses, the function never takes the value 1, so the range is all real numbers except 1: **R:** $-\infty < y < 1$ *or* $1 < y < \infty$.

43)   **What is the domain and the range of the following graph?**

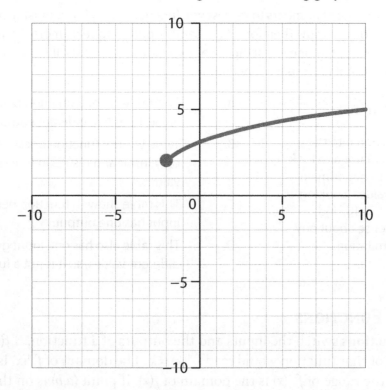

Answer:

For the domain, this graph goes on to the right to positive infinity. Its leftmost point, however, is $x = -2$. Therefore, its domain is all real numbers equal to or greater than −2, **D: − 2 ≤ x < ∞, or [−2, ∞).**

The lowest range value is $y = 2$. Although it has a decreasing slope, this function continues to rise. Therefore, the domain is all real numbers greater than 2, **R: 2 ≤ y < ∞ or [2, ∞).**

44)   **Which of the following represents a function?**

A.

x	g(x)
0	0
1	1
2	2
1	3

B.

x	f(x)
0	1
0	2
0	3
0	4

C.

t	f(t)
1	1
2	2
3	3
4	4

D.

x	f(x)
0	0
5	1
0	2
5	3

Answer:

For a set of numbers to represent a function, every input must generate a unique output. Therefore, if the same input (x) appears more than once in the table, determine if that input has two different outputs. If so, then the table does not represent a function.

A.   This table is not a function because input value 1 has two different outputs (1 and 3).

B.   Table B is not function because 0 is the only input and results in four different values.

C.   This table shows a function because each input has one output.

D.   This table also has one input going to two different values, so it is not a function.

INVERSE FUNCTIONS

Inverse functions switch the inputs and the outputs of a function. If $f(x) = k$ then the inverse of that function would read $f^{-1}(k) = x$. The domain of $f^{-1}(x)$ is the range of $f(x)$, and the range of $f^{-1}(x)$ is the domain of $f(x)$. If point (a,b) is on the graph of $f(x)$, then point (b,a) will be on the graph of $f^{-1}(x)$. Because of this fact, the graph of $f^{-1}(x)$ is a reflection of the graph of $f(x)$ across the line $y = x$. Inverse functions "undo" all the operations of the original function.

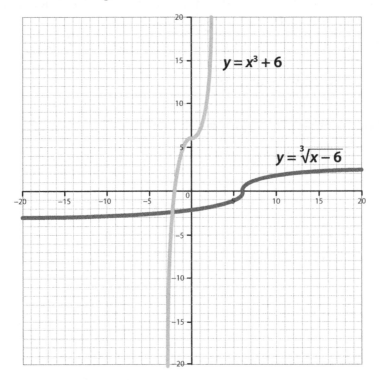

Figure 2.11. Inverse Functions

The steps for finding an inverse function are:

1. Replace $f(x)$ with y to make it easier manipulate the equation.

2. Switch the x and y.

3. Solve for y.

4. Label the inverse function as $f^{-1}(x) =$.

SAMPLE QUESTIONS

45) **What is the inverse of function of $f(x) = 5x + 5$?**

Answer:

$y = 5x + 5$	Replace $f(x)$ with y
$x = 5y + 5$	Switch the places of y and x.
$x = 5y + 5$ $x - 5 = 5y$ $y = \frac{x}{5} - 1$	Solve for y.
$f^{-1}(x) = \frac{x}{5} - 1$	

46) **Find the inverse of the graph of $f(x) = -1 - \frac{1}{5}x$**

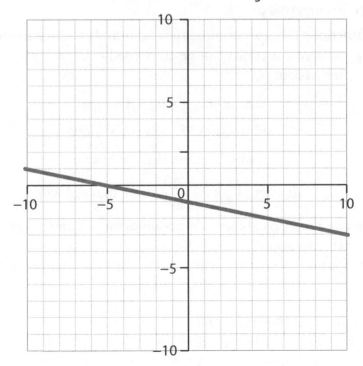

Answer:

This is a linear graph with some clear coordinates: $(-5, 0)$, $(0, -1)$, $(5, -2)$, and $(10, -3)$. This means the inverse function will have coordinate $(0, -5)$, $(-1, 0)$, $(-2, 5)$, and $(-3, 10)$. The inverse function is reflected over the line $y = x$ and is the line $f^{-1}(x) = -5(x + 1)$ below.

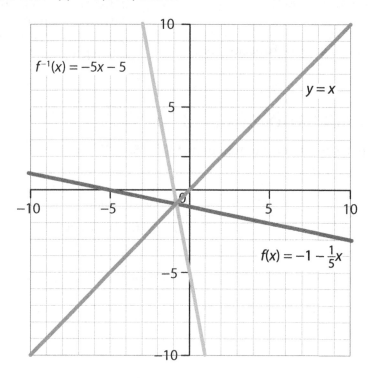

COMPOUND FUNCTIONS

Compound functions take two or more functions and combine them using operations or composition. Functions can be combined using addition, subtraction, multiplication, or division:

$$\text{addition: } (f + g)(x) = f(x) + g(x)$$

$$\text{subtraction: } (f - g)(x) = f(x) - g(x)$$

$$\text{multiplication: } (fg)(x) = f(x)g(x)$$

$$\text{division: } \left(\frac{f}{g}\right)(x) = \frac{f(x)}{g(x)} \text{ (note that } g(x) \neq 0)$$

Functions can also be combined using **composition**. Composition of functions is indicated by the notation $(f \circ g)(x)$. Note that the $\circ$ symbol does NOT mean multiply. It means take the output of $g(x)$ and make it the input of $f(x)$:

$$(f \circ g)(x) = f(g(x))$$

This equation is read f of g of x, and will be a new function of x. Note that order is important. In general, $f(g(x)) \neq g(f(x))$. They *will* be equal when $f(x)$ and $g(x)$ are inverses of each other, however, as both will simplify to the original input x. This is because performing a function on a value and then using that output as the input to the inverse function should bring you back to the original value.

The domain of a composition function is the set of x values that are in the domain of the "inside" function $g(x)$ such that $g(x)$ is in the domain of the outside function $f(x)$. For example, if $f(x) = \frac{1}{x}$ and $g(x) = \sqrt{x}$, $f(g(x))$ has a domain of $x > 0$ because $g(x)$ has a domain of $x \geq 0$. But when $f(x)$ is applied to the $\sqrt{x}$ function, the composition function becomes $\frac{1}{\sqrt{x}}$ and the value $x = 0$ is no longer allowed because it would result in 0 in the denominator, so the domain must be further restricted.

SAMPLE QUESTIONS

47) If $z(x) = 3x - 3$ and $y(x) = 2x - 1$, find $(y \circ z)(-4)$.

Answer:

$(y \circ z)(-4) = y(z(-4))$	
$z(-4)$ $= 3(-4) - 3$ $= -12 - 3$ $= -15$	Starting on the inside, evaluate z.
$y(z(-4))$ $= y(-15)$ $= 2(-15) - 1$ $= -30 - 1$ $= \mathbf{-31}$	Replace $z(-4)$ with -15, and simplify.

48) Find $(k \circ t)(x)$ if $k(x) = \frac{1}{2}x - 3$ and $t(x) = \frac{1}{2}x - 2$.

Answer:

$(k \circ t)(x) = k(t(x))$ $= k\left(\frac{1}{2}x - 2\right)$ $= \frac{1}{2}\left(\frac{1}{2}x - 2\right) - 3$	Replace x in the $k(x)$ function with $\left(\frac{1}{2}x - 2\right)$
$= \frac{1}{4}x - 1 - 3$ $= \frac{1}{4}x - 4$ $(k \circ t)(x) = \frac{1}{4}x - 4$	Simplify.

49) The wait (*W*) in minutes to get on a ride at an amusement park depends on the number of people (*N*) in the park. The number of people in the park depends on the number of hours, *t*, that the park has been open. Suppose $N(t) = 400t$ and $W(N) = 5(1.2)\frac{N}{100}$. What is the value and the meaning in context of $N(4)$ and $W(N(4))$?

Answer:

$N(4) = 400(4) = 1600$ and means that 4 hours after the park opens there are 1600 people in the park. $W(N(4)) = W(1600) = 96$ and means that 4 hours after the park opens the wait time is about **96 minutes** for the ride.

TRANSFORMING FUNCTIONS

Many functions can be graphed using simple transformation of parent functions. Transformations include reflections across axes, vertical and horizontal translations (or shifts), and vertical or horizontal stretches or compressions. The table gives the effect of each transformation to the graph of any function $y = f(x)$.

Table 2.4. Effects of Transformations

Equation	Effect on Graph
$y = -f(x)$	reflection across the x-axis (vertical reflection)
$y = f(x) + k$	vertical shift up k units (k > 0) or down k units (k < 0)
$y = kf(x)$	vertical stretch (if k > 1) or compression (if k < 1)
$y = f(-x)$	reflection across the y-axis (horizontal reflection)
$y = f(x + k)$	horizontal shift right k units (k < 0) or left k units (k > 0)
$y = f(kx)$	horizontal stretch (k < 1) or compression (k > 1)

Note that the first three equations have an operation applied to the *outside* of the function *f*(x) and these all cause *vertical changes* to the graph of the function that are *intuitive* (for example, adding a value moves it up). The last three equations have an operation applied to the *inside* of the function *f*(x) and these all cause *horizontal changes* to the graph of the function that are *counterintuitive* (for example, multiplying the x's by a fraction results in stretch, not compression, which would seem more intuitive). It is helpful to group these patterns together to remember how each transformation affects the graph.

SAMPLE QUESTIONS

50) **Graph:** $y = |x + 1| + 4$

Answer:

This function is the absolute value function with a vertical shift up of 4 units (since the 4 is outside the absolute value bars), and a horizontal shift left of 1 unit (since it is inside the bars). The vertex of the graph is at $(-1, 4)$ and the line $x = -1$ is an axis of symmetry.

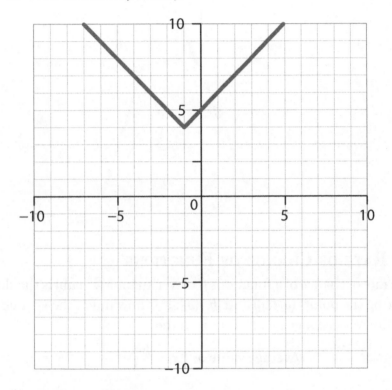

51) **Graph:** $y = -3|x - 2| + 2$

Answer:

The negative sign in front of the absolute value means the graph will be reflected across the x-axis, so it will open down. The 3 causes a vertical stretch of the function, which results in a narrower graph. The basic curve is shifted 2 units right (since the -2 is an inside change) and 2 units up (since the $+2$ is an outside change), so the vertex is at $(2, 2)$.

Go on →

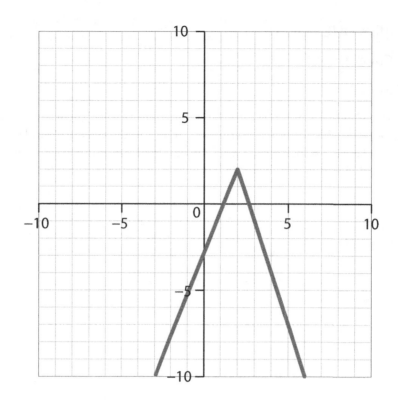

AVERAGE RATE OF CHANGE OF FUNCTIONS

The average rate of change of a functions $f(x)$ over an interval **[a,b]** is the slope of the line connecting the points $(a, f(a))$ and $(b, f(b))$. It is how much the y-value changes, on average, for every change of 1 unit in x on the interval:

$$\textbf{Average value of } \boldsymbol{f(x)} = \frac{f(b) - f(a)}{b - a}$$

SAMPLE QUESTION

52) Find the average rate of change of the function $f(x) = \sqrt{x-1}$ over the interval [5, 17].

Answer:

$\dfrac{f(b) - f(a)}{b - a}$	Apply the formula for the interval [5, 17].
$= \dfrac{f(17) - f(5)}{17 - 5}$	
$= \dfrac{\sqrt{17 - 1} - \sqrt{5 - 1}}{12}$	
$= \dfrac{4 - 2}{12}$	
$= \dfrac{2}{12}$	
$= \dfrac{1}{6}$	

EXPONENTIAL AND LOGARITHMIC FUNCTIONS

EXPONENTIAL FUNCTIONS

An **exponential function** has a constant base and a variable in the exponent: $f(x) = b^x$ is an exponential function with base b and exponent x. The value b is the quantity that the y value is multiplied by each time the x value is increased by 1. When looking at a table of values, an exponential function can be identified because the $f(x)$ values are being multiplied. (In contrast, linear $f(x)$ values are being added to.)

The graph of the exponential parent function does not cross the x-axis, which is the function's horizontal asymptote. The y-intercept of the function is at $(0,1)$.

The general formula for an exponential function, $f(x) = ab^{(x-h)} + k$, allows for transformations to be made to the function. The value h moves the function left or right (moving the y-intercept) while the value k moves the function up or down (moving both the y-intercept and the horizontal asymptote). The value a stretches or compresses the function (moving the y-intercept).

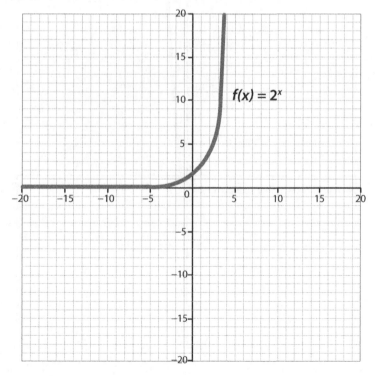

Figure 2.12. Exponential Parent Function

Exponential equations have at least one variable in an exponent position. One way to solve these equations is to make the bases on both side of the equation equivalent, and then equate the exponents. Many exponential equations do not have a solution. Negative numbers often lead to no solutions: for example, $2^x = -8$. The domain of exponential functions is only positive numbers, as seen above, so there is no x value that will result in a negative output.

SAMPLE QUESTIONS

53) Graph the exponential function $f(x) = 5^x - 2$.

Answer:

One way to do this is to use a table:

x	$5^x - 2$
−2	$\frac{1}{25} - 2 = -\frac{49}{25}$
−1	$\frac{1}{5} - 2 = -\frac{9}{5}$
0	$1 - 2 = -1$
1	$5 - 2 = 3$
2	$25 - 2 = 23$

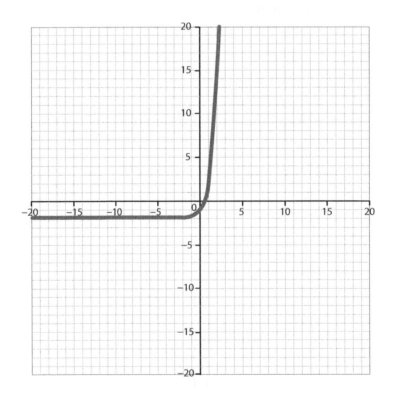

Another way to graph this is simply to see this function as the parent function $y = b^x$ (with $b = 5$), shifted down by a vertical shift of 2 units. Thus the new horizontal asymptote will be at $y = 2$, and the new y-intercept will be $y = -1$.

54) **If the height of grass in a yard in a humid summer week grows by 5% every day, how much taller would the grass be after six days?**

Answer:

Any time a question concerns growth or decay, an exponential function must be created to solve it. In this case, create a table with initial value a, and a daily growth rate of $(1+0.05) = 1.05$ per day.

Days (x)	Height (h)
0	a
1	$1.05a$
2	$1.05(1.05a) = (1.05)^2 a$
3	$(1.05)^2 (1.05a) = (1.05)^3 a$
x	$(1.05)^x a$

After six days the height of the grass is $(1.05)^6 = $ **1.34 times as tall**. The grass would grow 34% in one week.

55) **Solve for x: $4^{x+1} = \dfrac{1}{256}$**

Answer:

$4^{x+1} = \dfrac{1}{256}$	
$4^{x+1} = 4^{-4}$	Find a common base and rewrite the equation.
$x + 1 = -4$ $x = -5$	Set the exponents equal and solve for x.

LOGARITHMIC FUNCTIONS

The **logarithmic function (log)** is the inverse of the exponential function.

$$y = \log_3 x \Rightarrow 3^y = x$$

x	y
$\dfrac{1}{9}$	-2
$\dfrac{1}{3}$	-1

$$y = \log_3 x \Rightarrow 3^y = x \text{ (continued)}$$

1	0
3	1
9	2
27	3

A log is used to find out to what power an input is raised to get a desired output. In the table, the base is 3. The log function determines to what power 3 must be raised so that $\frac{1}{9}$ is the result in the table (the answer is -2). As with all inverse functions, these exponential and logarithmic functions are reflections of each other across the line $y = x$.

A **natural logarithm (ln)** has the number e as its base. Like π, e is an irrational number that is a nonterminating decimal. It is usually shortened to 2.71 when doing calculations. Although the proof of e is beyond the scope of this book, e is to be understood as the upper limit of the range of this rational function: $\left(1 + \frac{1}{n}\right)^n$.

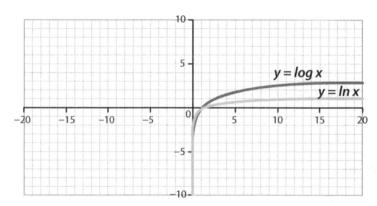

Figure 2.13. Logarithmic Parent Functions

In order to make use of and solve logarithmic functions, log rules are often employed that allow simplification:

Table 2.5. Properties of Logarithms

Change of base	$\log_b(m) = \dfrac{\log(m)}{\log(b)}$
Logs of products	$\log_b(mn) = \log_b(m) + \log_b(n)$
Logs of quotients	$\log_b\left(\frac{m}{n}\right) = \log_b(m) - \log_b(n)$
Log of a power	$\log_b(m^n) = n \times \log_b(m)$
Equal logs/equal arguments	$\log_b M = \log_b N \Leftrightarrow M = N$

Note that when the base is not written out, such as in log(m), it is understood that the base is 10. Just like a 1 is not put in front of a variable because its presence is implicitly understood, 10 is the implicit base whenever a base is not written out.

SAMPLE QUESTIONS

56) **Expand** $\log_5\left(\frac{25}{x}\right)$

Answer:

Since division of a term can be written as a subtraction problem, this simplifies to:

$\log_5(25) - \log_5(x)$

The first term asks "what power of 5 gives 25?" The power is 2. Therefore, the most expanded form is:

2 − $\log_5(x)$

57) **Solve for x: $\ln x + \ln 4 = 2\ln 4 - \ln 2$**

Answer:

$\ln x + \ln 4 = 2\ln 4 - \ln 2$	
$\ln(4x) = \ln 4^2 - \ln 2$ $\ln(4x) = \ln 16 - \ln 2$	Apply the log of product and log of exponent rules.
$\ln(4x) = \ln 8$	Follow log of quotient rule.
$4x = 8$ $x = 2$	Set the arguments equal to each other.

58) **Solve for x: $2^x = 40$**

Answer:

$\log_2 2^x = \log_2 40$	Take the $\log_2$ of both sides.
$x\log_2 2 = \log_2 40$	Drop the x down using properties of logs.
$x = \log_2 40$	$\log_2 2$ simplifies to 1.
≈ 5.32	Use the change of base rule or a calculator to calculate the value of $\log_2(40)$.

SPECIAL EQUATIONS

There are three exponential function formulas that frequently show up in word problems:

The **Growth Formula:**

$y = a(1 + r)^t$ Initial amount a increases at a rate of r per time period

The **Decay Formula:**

$y = a(1 - r)^t$ Initial amount a decreases at a rate of r per time period

In these formulas, a is the initial amount (at time $t = 0$), r is the rate of growth or decay (written as a decimal in the formula), and t is the number of growth or decay cycles that have passed.

A special case of the growth function is known as the **Compound-Interest Formula**:

$$A = P\left(1 + \frac{r}{n}\right)^{nt}$$

In this formula, A is the future value of an investment, P is the initial deposit (or principal), r is the interest rate as a percentage, n is the number of times interest is compounded within a time period, or how often interest is applied to the account in a year (once per year, $n = 1$; monthly, $n = 12$; etc.), and t is the number of compounding cycles (usually years).

SAMPLE QUESTIONS

59) **In the year 2000, the number of text messages sent in a small town was 120. If the number of text messages grew every year afterward by 124%, how many years would it take for the number of text messages to surpass 36,000?**

    Answer:

$y = a(1 + r)^t$ $36{,}000 = 120(1 + 1.24)^t$	Plug the given values into the growth equation.
$300 = (2.24)^t$ $\log_{2.24} 300 = \log_{2.24}(2.24)^t$ $7.07 \approx t$ The number of text messages will pass 36,000 in **7.07 years**.	Use the properties of logarithms to solve the equation.

60) **The half-life of a certain isotope is 5.5 years. If there were 20 grams of one such isotope left after 22 years, what was its original weight?**

Answer:

$t = \frac{22}{5.5} = 4$ $r = 0.5$ $a = ?$	Identify the variables.
$20 = a(1 - 0.50)^4$ $20 = a(0.5)^4$ $20 = a(\frac{1}{2})^4$ $20 = a(\frac{1}{16})$ $320 = a$ The original weight is **320 grams**.	Plug these values into the decay formula and solve.

61) **If there were a glitch at a bank and a savings account accrued 5% interest five times per week, what would be the amount earned on a $50 deposit after twelve weeks?**

Answer:

$r = 0.05$ $n = 5$ $t = 12$ $P = 50$	Identify the variables.
$A = 50 \left(1 + \frac{0.05}{5}\right)^{5(12)}$ $A = 50(1.01)^{60}$ $A = 50(1.82) = 90.83$	Use the compound-interest formula, since this problem has many steps of growth within a time period.
$90.83 - 50$ $= \mathbf{\$40.83}$	Subtract the original deposit to find the amount of interest earned.

POLYNOMIAL FUNCTIONS

A polynomial is any equation or expression with two or more terms with whole number exponents. All polynomials with only one variable are functions. The zeros, or roots, of a polynomial function are where the function equals zero and crosses the x-axis.

A linear function is a degree 1 polynomial and always has one zero. A quadratic function is a degree 2 polynomial and always has exactly two roots (including complex roots and counting repeated roots separately). This pattern is extended in the **Fundamental Theorem of Algebra:**

A polynomial function with degree $n > 0$ such as $f(x) = ax^n + bx^{n-1} + cx^{n-2} + \ldots + k$, has exactly n (real or complex) roots (some roots may be repeated). Simply stated, whatever the degree of the polynomial is, that is how many roots it will have.

Table 2.6. Zeros of Polynomial Functions

Polynomial Degree, n	Number and Possible Types of Zeros
1	1 real zero (guaranteed)
2	0, 1, or 2 real zeros possible
	2 real **or** complex zeros (guaranteed)
3	1, 2, or 3 real zeros possible (there must be at least one real zero)
	Or 1 real zero (guaranteed) and 2 complex zeros (guaranteed)
4	0, 1, 2, 3, or 4 real zeros (possible)
	Or 2 real zeros and 2 complex zeros or 4 complex zeros
…	…

QUICK REVIEW

All polynomials where n is an odd number will have at least one real zero or root. Complex zeros always come in pairs (specifically, complex conjugate pairs).

All the zeros of a polynomial satisfy the equation $f(x) = 0$. That is, if k is a zero of a polynomial, then plugging in $x = k$ into the polynomial results in 0. This also means that the polynomial is evenly divisible by the factor $(x - k)$.

SAMPLE QUESTION

62) Find the roots of the polynomial: $y = 3t^4 - 48$

Answer:

$y = 3t^4 - 48$	
$3(t^4 - 16) = 0$	Factor the polynomial. Remove the common factor of 3 from each term and make $y = 0$.
$3(t^2 - 4)(t^2 + 4) = 0$ $3(t + 2)(t - 2)(t^2 + 2) = 0$	Factor the difference of squares. $t^2 - 4$ is also a difference of squares.
$t + 2 = 0 \quad t - 2 = 0 \quad t^2 + 2 = 0$ $t = -2 \qquad t = 2 \qquad t^2 = -2$ $\qquad\qquad\qquad\qquad t = \pm\sqrt{-2} = \pm 2i$	Set each factor equal to zero. Solve each equation.

This degree 4 polynomial has four roots, two real roots: **2 or –2**, and two complex roots: **2i or –2i**. The graph will have two *x*-intercepts at $(-2, 0)$ and $(2, 0)$.

RATIONAL FUNCTIONS

OPERATIONS WITH RATIONAL FUNCTIONS

Rational functions are ratios of polynomial functions in the form $f(x) = \frac{g(x)}{h(x)}$. Just like rational numbers, rational functions form a closed system under addition, subtraction, multiplication, and division by a nonzero rational expression. This means adding two rational functions, for example, results in another rational function.

To add or subtract rational expressions, the least common denominator of the factors in the denominator must be found. Then, numerators are added, just like adding rational numbers. To multiply rational expressions, factors can be multiplied straight across, canceling factors that appear in the numerator and denominator. To divide rational functions, use the "invert and multiply" rule.

Rational equations are solved by multiplying through the equation by the least common denominator of factors in the denominator. Just like with radical equations, this process can result in extraneous solutions, so all answers need to be checked by plugging them into the original equation.

SAMPLE QUESTIONS

63) If $f(x) = \frac{2}{3x^2y}$ and $g(x) = \frac{5}{21y}$, find the difference between the functions, $f(x) - g(x)$.

Answer:

$f(x) - g(x) = \frac{2}{3x^2y} - \frac{5}{21y}$	Write the difference.
$= \frac{2}{3x^2y}\left(\frac{7}{7}\right) - \frac{5}{21y}\left(\frac{x^2}{x^2}\right)$ $= \frac{14}{21x^2y} - \frac{5x^2}{21x^2y}$	Figure out the least common denominator. Every factor must be represented to the highest power it appears in either denominator. So, the LCD $= 3(7)x^2y$.
$f(x) - g(x) = \frac{14 - 5x^2}{21x^2y}$	Subtract the numerators the find the answer.

64) If $f(x) = \frac{(x-1)(x+2)^2}{5x^2 + 10x}$ and $g(x) = \frac{x^2 + x - 2}{x + 5}$, find the quotient $\frac{f(x)}{g(x)}$.

Answer:

$\dfrac{f(x)}{g(x)} = \dfrac{\dfrac{(x-1)(x+2)^2}{5x^2+10x}}{\dfrac{x^2+x-2}{x+5}}$ $= \dfrac{(x-1)(x+2)^2}{5x^2+10x} \times \dfrac{x+5}{x^2+x-2}$	Write the quotient; then invert and multiply.
$= \dfrac{(x-1)(x+2)^2}{5x(x+2)} \times \dfrac{x+5}{(x+2)(x-1)}$	Factor all expressions, and then cancel any factors that appear in both the numerator and the denominator.
$= \dfrac{x+5}{5x}$	

65) **Solve the rational equation** $\dfrac{x}{x+2} + \dfrac{2}{x^2+5x+6} = \dfrac{5}{x+3}$.

Answer:

$\dfrac{x}{x+2} + \dfrac{2}{x^2+5x+6} = \dfrac{5}{x+3}$	
$\dfrac{x}{x+2} + \dfrac{2}{(x+3)(x+2)} = \dfrac{5}{x+3}$	Factor any denominators that need factoring.
$x(x+3) + 2 = 5(x+2)$	Multiply through by the LCM of the denominators, which is $(x+2)(x+3)$.
$x^2 + 3x + 2 - 5x - 10 = 0$ $x^2 - 2x - 8 = 0$	Simplify the expression.
$(x-4)(x+2) = 0$	Factor the quadratic.

Plugging $x = -2$ into the original equation results in a 0 in the denominator. So this solution is an extraneous solution and must be thrown out.

Plugging in $x = 4$ gives $\dfrac{4}{6} + \dfrac{2}{16+20+6} = \dfrac{5}{7}$.

So $x = 4$ is a solution to the equation.

GRAPHING RATIONAL FUNCTIONS

Rations functions are graphed by examining the function to find key features of the graph, including asymptotes, intercepts, and holes.

A **vertical asymptote** exists at any value that makes the denominator of a (simplified) rational function equal zero. A vertical asymptote is a vertical line through an x value that is not in the domain of the rational function (the function is undefined at this value because division by 0 is not allowed). The function approaches, but never crosses, this line, and the y values increase (or decrease) without bound (or "go to infinity") as this x value is approached.

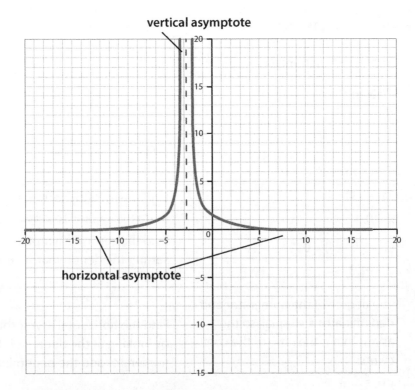

Figure 2.14. Graphing Rational Functions

To find *x*-intercepts and vertical asymptotes, factor the numerator and denominator of the function. Cancel any terms that appear in the numerator and denominator (if there are any). These values will appear as **holes** on the final graph. Since a fraction only equals 0 when its numerator is 0, set the simplified numerator equal to 0 and solve to find the *x*-intercepts. Next, set the denominator equal to 0 and solve to find the vertical asymptotes.

Horizontal asymptotes are horizontal lines that describe the "end behavior" of a rational function. In other words, the horizontal asymptote describes what happens to the *y*-values of the function as the *x*-values get very large ($x \to \infty$) or very small ($x \to -\infty$). A horizontal asymptote occurs if the degree of the numerator of a rational function is less than or equal to the degree in the denominator. The table summarizes the conditions for horizontal asymptotes:

Table 2.7. Conditions for Horizontal Asymptotes

For polynomials with first terms $\frac{ax^n}{bx^d}$ ...

$n < d$	as $x \to \infty$, $y \to 0$ as $x \to -\infty$, $y \to 0$	The *x*-axis ($y = 0$) is a horizontal asymptote.
$n = d$	as $x \to \pm\infty$, $y \to \frac{a}{b}$	There is a horizontal asymptote at $y = \frac{a}{b}$.
$n > d$	as $x \to \infty$, $y \to \infty$ or $-\infty$ as $x \to -\infty$, $y \to \infty$ or $-\infty$	There is no horizontal asymptote.

SAMPLE QUESTIONS

66) **Graph the function:** $f(x) = \frac{3x^2 - 12x}{x^2 - 2x - 3}$.

Answer:

$y = \frac{3x^2 - 12x}{x^2 - 2x - 3}$ $= \frac{3x(x-4)}{(x-3)(x+1)}$	Factor the equation.
$3x(x-4) = 0$ $x = 0, 4$	Find the roots by setting the numerator equal to zero.
$(x-3)(x+1) = 0$ $x = -1, 3$	Find the vertical asymptotes by setting the denominator equal to zero.
The degree of the numerator and denominator are equal, so the asymptote is the ratio of the coefficients: $y = \frac{3}{1} = 3$	Find the horizontal asymptote by looking at the degree of the numerator and the denominator.

Use the roots and asymptotes to graph the function.

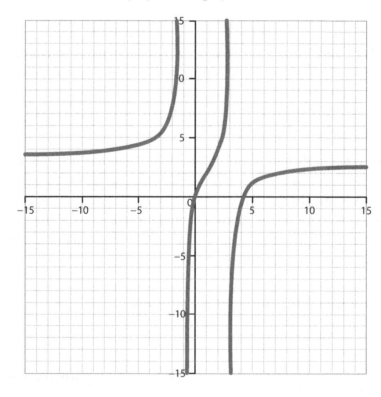

67) **Create a function that has an *x*-intercept at (5, 0) and vertical asymptotes at *x* = 1 and *x* = −1.**

Answer:

The numerator will have a factor of $(x - 5)$ in order to have a zero at $x = 5$. The denominator will need factors of $(x - 1)$ and $(x + 1)$ in order for the denominator to be 0 when x is 1 or –1. Thus, one function that would have these features is

$$y = \frac{(x-5)}{(x+1)(x-1)} = \frac{x-5}{x^2-1}$$

RADICAL FUNCTIONS

Radical functions have rational (fractional) exponents, or include the radical symbol. For example, $f(x) = 2(x - 5)^{\frac{1}{3}}$ and $g(t) = \sqrt[4]{5 - x}$ are radical functions. The domain of even root parent functions is $0 \leq x \leq \infty$ and the range is $y \geq 0$. For odd root parent functions, the domain is all real numbers (because you can take cube roots, etc., of negative numbers). The range is also all real numbers.

To solve equations involving radical functions, first isolate the radical part of the expression. Then "undo" the fractional exponent by raising both sides to the reciprocal of the fractional exponent (for example, undo square roots by squaring both sides). Then solve the equation using inverse operations, as always. All answers should be checked by plugging them back into the original equation, as **extraneous solutions** result when an equation is raised to powers on both sides. This means there may be some answers that are not actually solutions, and should be eliminated.

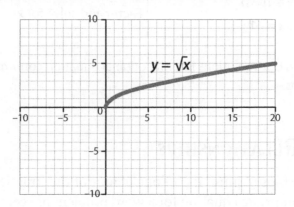

Figure 2.15. Radical Parent Function

SAMPLE QUESTIONS

68) **Solve the equation:** $\sqrt{2x - 5} + 4 = x$

Answer:

$\sqrt{2x - 5} + 4 = x$	
$\sqrt{2x - 5} = x - 4$	Isolate the $\sqrt{2x - 5}$ by subtracting 4.

$2x - 5 = x^2 - 8x + 16$	Square both sides to clear the $\sqrt{\ }$.
$x^2 - 10x + 21 = 0$	Collect all variables to one side.
$(x - 7)(x - 3) = 0$ $x = 7$ or $x = 3$	Factor and solve.
$\sqrt{2(7) - 5} + 4 = 7$ $\sqrt{2(3) - 5} + 4 = 3$ $\sqrt{9} + 4 = 7$ $\sqrt{1} + 4 = 3$ $x = 7$	Check solutions by plugging into the original, as squaring both sides can cause extraneous solutions. True, $x = 7$ is a solution. False, $x = 3$ is NOT a solution (extraneous solution).

69) Solve the equation: $2(x^2 - 7x)^{\frac{2}{3}} = 8$

Answer:

$2(x^2 - 7x)^{\frac{2}{3}} = 8$	
$(x^2 - 7x)^{\frac{2}{3}} = 4$	Divide by 2 to isolate the radical.
$x^2 - 7x = 4^{\frac{3}{2}}$ $x^2 - 7x = 8$	Raise both sides to the $\frac{3}{2}$ power to clear the $\frac{2}{3}$ exponent.
$x^2 - 7x - 8 = 0$	This is a quadratic, so collect all terms to one side.
$(x - 8)(x + 1) = 0$ $x = 8$ or $x = -1$	Factor and solve for x.

Plugging both solutions into the original equation confirms that both are solutions.

MODELING RELATIONSHIPS

Modeling relationships requires use of one of four of the function types examined above with an appropriate equation for a word problem or scenario.

Table 2.8. Function Types

Linear

$y = x$

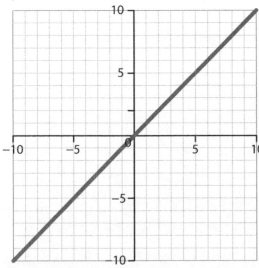

Key words: constant change, slope, equal

Quadratic

$y = x^2$

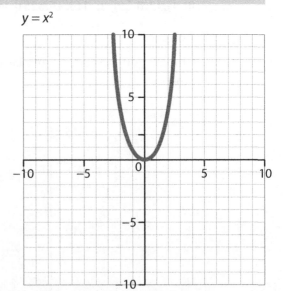

Key words: area, squared, parabola

Exponential

$y = a^x$

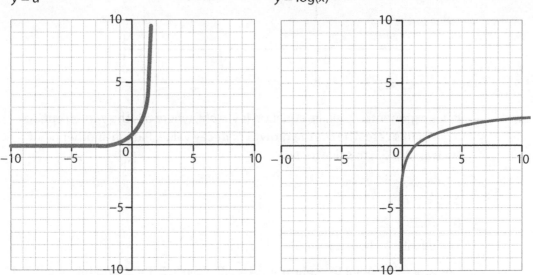

Key words: growth, decay, interest, double, triple, half-life

Logarithmic

$y = \log(x)$

Key words: log-scale, base, log equations

Since exponential functions and log functions are inverses of each other, it will often be the case that exponential or log problems can be solved by either type of equation.

SAMPLE QUESTIONS

70) **Consider the following sets of coordinate pairs of a function: {(-1, 0.4), (0, 1), (2, 6.25), and (3, 15.625)}. What kind of function does this represent?**

Answer:

Graphing on the coordiante plane shows what looks like an exponential function.

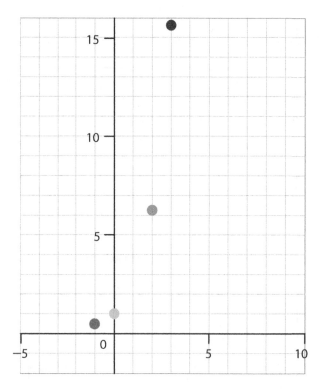

If it is exponential, then its equation is $y = ab^x$, where a is the y-intercept, so $a = 1$ in this case. The b is the growth or decay value. Plug in another point, such as (2, 6.25) to solve for b:

$y = ab^x$

$6.25 = (1)b^2$

$b = \sqrt{6.25} = 2.5$

The equation, then, is $y = 2.5^x$.

Check another point to confirm: Is $0.4 = 2.5^{-1}$? Since $2.5 = \frac{5}{2}$, and $\left(\frac{5}{2}\right)^{-1} = \frac{2}{5} = 0.4$, the equation works. The function is **exponential**.

71) **At a recent sporting event, there were 20,000 people in attendance. When it ended, people left the building at a rate of 1,000 people in the first minute, 1,000 more in the second minute, 1,000 in the third minute, and so on. What equation describes the behavior of attendees leaving the event for every minute after the event finished?**

Answer:

The dependent variable is the number of attendees leaving the event (y). There is a constant change of 1,000 people per minute. Note that this is an additive pattern in the table: every increase of 1 in time results in a subtraction of the same value (1,000) in y. Because it is a constant rate of change, a linear model is required:

$y = 20,000 - 1,000x$

Here 20,000 is the y-intercept, and the rate of change, −1,000, is the slope.

To test this model, confirm that 18,000 attendees were left in the building after two minutes:

$y = 20,000 - 1,000(2) = 18,000$

The model is correct.

TRIGONOMETRY

Trigonometry comes from the Greek words for *triangle* and *measuring*. Appropriately enough, trigonometry is used to find missing angles or side lengths in a triangle. Trigonometric questions often require use of algebraic skills with geometric concepts.

THE SIX TRIGONOMETRIC FUNCTIONS

There are six different trigonometric functions that are the foundations of trigonometry. They can be thought of as three pairs, as they are reciprocals of one another. All of these functions are ratios of the side lengths of a right triangle. The longest side of a right triangle (opposite the 90-degree angle) is called the **hypotenuse**. The side directly opposite the angle being used is the **opposite**, and the side next to the angle is called the **adjacent** side.

> **HELPFUL HINT**
>
> SOHCAHTOA, or Some Old Horse Caught Another Horse Taking Oats Away, is a way to remember that Sine is Opposite over Hypotenuse, Cosine is Adjacent over Hypotenuse, and Tangent is Opposite over Adjacent.

Table 2.9. The Six Trigonometric Functions

Sine Function	Cosine Function	Tangent Function
$\sin\theta = \dfrac{\text{opposite}}{\text{hypotenuse}}$	$\cos\theta = \dfrac{\text{adjacent}}{\text{hypotenuse}}$	$\tan\theta = \dfrac{\text{opposite}}{\text{adjacent}}$
Cosecant Function	Secant Function	Cotangent Function
$\csc\theta = \dfrac{\text{hypotenuse}}{\text{opposite}}$	$\sec\theta = \dfrac{\text{hypotenuse}}{\text{adjacent}}$	$\cot\theta = \dfrac{\text{adjacent}}{\text{opposite}}$

There are a couple of special triangles that are used frequently and whose properties are worth memorizing. They are the 30° – 60° – 90° and the 45° – 45° – 90° triangles.

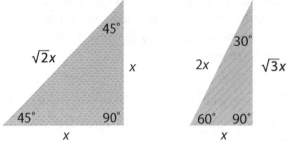

Figure 2.16. Special Right Triangles

For the 30° – 60° – 90° triangle, the side opposite the 30° angle (the shortest side) is always half the hypotenuse length, and the medium side (opposite the 60° angle) is $\sqrt{3}$ times the shortest side.

For the 45° – 45° – 90° triangle, two legs have the same length since two angles are equal (the triangle is isosceles), and the hypotenuse is always $\sqrt{2}$ times the length of a leg.

SAMPLE QUESTIONS

72) **Find all six trigonometric ratios for θ in the following triangle:**

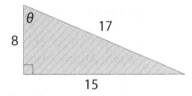

Answer:

The side directly opposite of θ is 15. The hypotenuse is the longest side with a length of 17. This leaves the 8 as the adjacent.

Sine Function	Cosine Function	Tangent Function
$\sin\theta = \frac{15}{17} = 0.88$	$\cos\theta = \frac{8}{17} = 0.47$	$\tan\theta = \frac{15}{8} = 1.88$
Cosecant Function	Secant Function	Cotangent Function
$\csc\theta = \frac{17}{15} = 1.13$	$\sec\theta = \frac{17}{8} = 2.13$	$\cot\theta = \frac{8}{15} = 0.53$

73) **Find the missing length:**

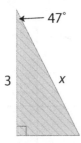

Answer:

Identify which side length is known, and which is being solved for in relation to the given angle. With respect to the 47-degree angle, the **h**ypotenuse is the unknown, and the known value is the **a**djacent. The trig function that uses adjacent and hypotenuse is cosine.

$\theta = 47$ degrees	
$a = 3$	Identify the given parts of the triangle.
$h = x$	
$\cos 47° = \frac{3}{x}$	
$x(\cos 47°) = 3$	Plug these values into the equation for
$x = \frac{3}{\cos 47°}$	cosine and solve.
$= \mathbf{4.40}$	

74) **Find the angle θ in degrees:**

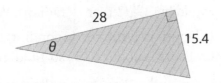

Answer:

$adjacent = 28$	
$opposite = 15.4$	Identify the given parts of the
$\theta = ?$	triangle.
$\tan \theta = \frac{opposite}{adjacent}$	
$\tan \theta = \frac{15.4}{28}$	Use the equation for the tangent
$\theta = \tan^{-1}\left(\frac{15.4}{28}\right)$	function to find the angle.
$\boldsymbol{\theta = 28.81°}$	

The Unit Circle

The **unit circle** is on the coordinate plane, with its center at the origin (0,0) and a radius of 1. By using triangles within the unit circle (which will all have a hypotenuse of 1), the trigonometric ratios can be extended so that trigonometric functions of *any* angle may be evaluated. In fact, each point (*x,y*) on the unit circle can be expressed as trig functions of the angle θ: (*x,y*) = (cosθ, sinθ).

An angle in **standard position** in the plane has an initial (beginning) ray on the *x*-axis and a terminal (end) ray on the radius of the circle. Positive angles are measured in the counterclockwise direction, and negative angles are measured clockwise from the *x*-axis. One complete circle contains 360°.

Another way to measure angles is with radians, which involves finding the length of the arc on the circle intercepted by the terminal ray of the angle. Since the circumference of the circle is $C = 2\pi r = 2\pi(1) = 2\pi$, the angle corresponding to one complete circle (360°) has a radian measure of 2π. Other angles can be expressed as fractions of 2π. For example, 90° is $\frac{1}{4}$ of a circle, so its radian measure is $\frac{1}{4}(2\pi) = \frac{\pi}{2}$. A 30° angle would be $\frac{30}{360}(2\pi)$ or $\frac{\pi}{6}$ radians. When the angle intersects the circle such that the arc length is 1, the corresponding angle is 1 radian. The angle in degrees at which this occurs is about 57.3°, so 1 radian ≈ 57.3°.

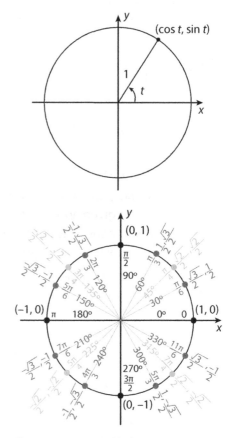

Figure 2.17. Unit Circle

Angles can be converted between degrees and radians using these conversion factors:

$$\text{degrees} = \text{radians} \times \frac{180}{\pi} \qquad\qquad \text{radians} = \text{degrees} \times \frac{\pi}{180}$$

Any angle has an infinite number of **coterminal angles** associated with it. These are angles that share the same terminal ray. For example, 390° is coterminal with 30°, because 390° is one complete revolution around the circle plus 30° more, so it lands on the same terminal ray. Another co-terminal angle to this angle would be –330°. To find co-terminal angles in degrees, simply add to or subtract 360° from the angle. In radians, add to or subtract 2π from the angle.

HELPFUL HINT

A calculator can be put in either radian or degree mode as appropriate for the given problem.

The unit circle diagram is made up of a number of specific sine and cosine coordinates for angles that are frequently used (often called special angles). Tangent on the unit circle is defined as the ratio of sine to cosine, $\tan\theta = \frac{\sin\theta}{\cos\theta}$.

Table 2.10. Special Angle Values in the First Quadrant

Degrees	0	30	45	60	90
Radians	0	$\frac{\pi}{6}$	$\frac{\pi}{4}$	$\frac{\pi}{3}$	$\frac{\pi}{2}$
$\sin\theta$	0	$\frac{1}{2}$	$\frac{\sqrt{2}}{2}$	$\frac{\sqrt{3}}{2}$	1
$\cos\theta$	1	$\frac{\sqrt{3}}{2}$	$\frac{\sqrt{2}}{2}$	$\frac{1}{2}$	0
$\tan\theta$	0	$\frac{\sqrt{3}}{3}$	1	$\sqrt{3}$	undefined

When calculating trig functions of angles in other quadrants, make a sketch of the angle and drop a perpendicular altitude down to the nearest x-axis. This forms a triangle. The angle between the x-axis and the terminal ray is called the **reference angle.** It will always be an angle between 0 and $\frac{\pi}{2}$ (or 0 and 90°). If it is one of the special angles, either label the sides of the triangles using the special triangle rules, or use the table above to find the value. In either case, care must be given to the *sign* of the value. As the terminal ray travels into quadrants 2, 3, and 4, the signs of the x- and y-coordinates are sometimes negative, so the corresponding trig functions will also be negative. This diagram summarizes where each trig function is *positive* (where it is not positive, it is negative!).

> **HELPFUL HINT**
>
> To remember in which quadrant each trig function is positive, starting in quad 1, remember **ASTC**, or **A**ll **S**tudents **T**hrow **C**halk: A = all, S = sine, T = tangent, and C = cosine.

Q 2 $\sin\theta$ and $\csc\theta$ +	Q 1 ALL trig functions +
Q 3 $\tan\theta$ and $\cot\theta$ +	Q 4 $\cos\theta$ and $\sec\theta$ +

Figure 2.18. Trigonometric Signs by Quadrant

SAMPLE QUESTIONS

75) Find $\sin\frac{\pi}{2}$.

   Answer:

   To make use of a graphing calculator's trigonometric function, make sure it is in radian mode and type in: $\sin\frac{\pi}{2}$, which returns a value of 1. To understand *why* this is true, locate the angle $\frac{\pi}{2}$ on the unit circle. The $\sin\frac{\pi}{2}$ is the y-coordinate of the intersection of the terminal ray of angle $\frac{\pi}{2}$ with the unit circle, which is 1 because the ray intersects the circle at point (0, 1).

76) Find $\csc\left(\frac{7\pi}{4}\right)$.

   Answer:

$\csc\frac{7\pi}{4} = \dfrac{1}{\sin\frac{7\pi}{4}}$	Rewrite the expression using the reciprocal identity.
$\sin\frac{7\pi}{4} = \dfrac{-1}{\sqrt{2}}$	Find $\frac{7\pi}{4}$ on the unit circle.
$\csc\frac{7\pi}{4} = \dfrac{1}{\sin\frac{7\pi}{4}} = \dfrac{1}{\frac{-1}{\sqrt{2}}} = -\sqrt{2}$	Convert sine into cosecant.

GRAPHING TRIGONOMETRIC FUNCTIONS

Each trigonometric function can be graphed as a function of the angle θ. The graph of $\sin\theta$ can be understood by considering the height of the altitude of the right triangle (the y value) in the unit circle at different values of θ. Consider what happens to the y values as θ increases as it goes around the circle from 0 to 2π: $\sin\theta$ begins at 0, because there is no height when the angle is 0; then the height increases to a maximum value of 1 as θ increases to $\frac{\pi}{2}$. Then it decreases back to 0 at $\theta = \pi$, and so on. Similarly, the x-values associated with the triangle within the unit circle begin at 1 and vary between 1 and –1, which traces out the $\cos\theta$ curve. Both the sine and cosine functions trace out waves and are called **sinusoidal** graphs/functions.

Except sine and cosine, what defines all these graphs are the many asymptotes. These occur when the ratios that make up the trigonometric functions have a zero in the denominator. For example, since $\csc\theta = \frac{1}{\sin\theta}$, whenever $\sin\theta$ is equal to zero, $\csc\theta$ has an asymptote at $\theta = 0$.

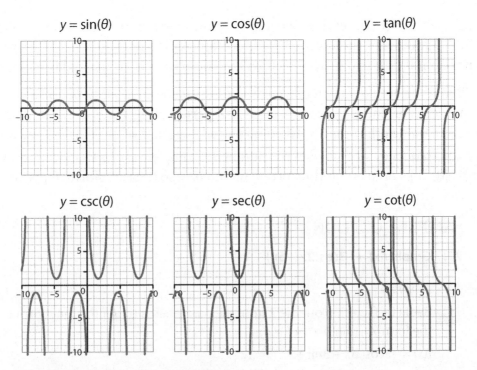

Figure 2.19. Graphs of Trigonometric Functions

Note that all of the trigonometric functions are **periodic.** This means they repeat at regular intervals called the **period.** The period of the functions $y = \sin\theta$, $\cos\theta$, $\csc\theta$, and $\sec\theta$ is 2π. (This is because each revolution around the unit circle traces out one complete cycle of the graph.) The functions $y = \tan\theta$ and $\cot\theta$ both have a period of π.

The transformations learned in this chapter can be applied to these new trigonometric parent functions. There is some specialized vocabulary associated with these functions, however. Consider the function $y = a\sin(b(x + c)) + d$ (sin could be replaced with cos). For sine and cosine, the **midline** of the graph is the value of the vertical shift (d). It is the horizontal line through the middle of the wave. The **amplitude** of the wave is the vertical stretch or compression value ($|a|$). It gives

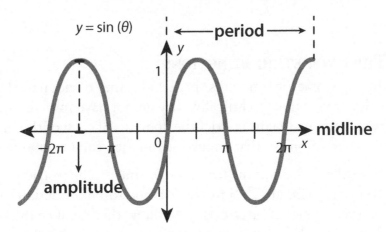

Figure 2.20. Parts of a Trigonometric Graph

the distance the graph reaches above and below the midline (amplitude is always positive). As always, if the a is negative, the graph will be reflected across the x-axis. The horizontal shift is sometimes called the **phase shift** in sinusoidal functions (c).

Finally, the horizontal stretch or compression value (b) causes a period change in the function. For sine and cosine, the normal period of the function is 2π, so when there is a b-value other than 1, the new period is period = $\frac{2\pi}{b}$. The value b is also called the **frequency** since it gives the number of complete cycles that occur over the interval $[0, 2\pi]$. For tangent, the normal period is π, so the transformed period is $\frac{\pi}{b}$.

SAMPLE QUESTION

77)   Graph: $f(x) = -3\cos(2x - 6\pi) + 1$

Answer:

Begin by factoring out the 2 from the parentheses to get the function into the form $y = a\sin(b(x + c)) + d$:

$f(x) = -3\cos(2(x - 3\pi)) + 1$

The a value of −3 means the graph wil be reflected across the x-axis, and the amplitude of the graph is 3. The b value of 2 means the period is $\frac{2\pi}{2} = \pi$. The c value of −3π means the graph will be shifted to the right 3π units. The d value of 1 indicates a vertical translation up 1 unit.

$f(x) = -3\cos(2x - 6\pi) + 1$

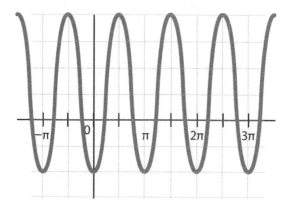

INVERSE TRIGONOMETRIC FUNCTIONS

Because all the trigonometric functions are periodic, they do not pass the horizontal line test. This means that, technically, they do not have inverse functions. In reality, however, mathematicians restrict the domains of each of the functions to force them to be one-to-one so that an inverse function can be obtained.

For instance, the domain of the function $y = \sin\theta$ can be restricted (as shown in Figure 2.21) to the interval $[\frac{-\pi}{2}, \frac{\pi}{2}]$ to create a one-to-one function that is then invertible. The inverse function, **arcsin(y)** or $\theta = \sin^{-1}y$, then would have input values (or a domain) of $-1 \le y \le 1$, and output values (or a range) of $\frac{-\pi}{2} \le \theta \le \frac{\pi}{2}$.

Similarly, the tangent would be restricted in the same way, but the input values for the **arctan(*y*)** function would be any real number (since the range of the tangent function is all real numbers). The cosine requires a different restriction to make it one-to-one. The restriction for cosine is the interval **[0,π]**, so the domain of **arccos(*y*)** is –1 ≤ *y* ≤ 1, and the range is **[0,π]**.

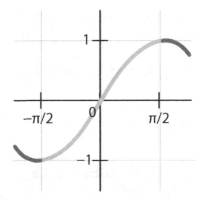

Figure 2.21. $y = \sin\theta$

The inverse trig functions on the calculator employ these same restrictions. This is why, when **arcsin($\frac{-1}{2}$)** is plugged in, an answer equivalent to $\frac{-\pi}{6}$ results (with the calculator in radian mode). While getting a negative answer may seem strange, it is because of the restrictions on the domain and range that this occurs.

Table 2.11. Standard Restricted Domains

Trigonometric Function	Domain	Range
sin$x$	$\left[\frac{-\pi}{2}, \frac{\pi}{2}\right]$	$[-1, 1]$
cos$x$	$[0, \pi]$	$[-1, 1]$
tan$x$	$\left[\frac{-\pi}{2}, \frac{\pi}{2}\right]$	$(-\infty, \infty)$
csc$x$	$[0, \pi]$	$(-\infty, \infty)$
sec$x$	$\left[0, \frac{\pi}{2}\right) \cup \left(\frac{\pi}{2}, \pi\right]$	$(-\infty, -1) \cup [1, \infty)$
cot$x$	$\left[\frac{-\pi}{2}, 0\right) \cup \left(0, \frac{\pi}{2}\right]$	$(-\infty, -1) \cup [1, \infty)$
sin^{-1}x	$[-1, 1]$	$\left[\frac{-\pi}{2}, \frac{\pi}{2}\right]$
cos^{-1}x	$[-1, 1]$	$[0, \pi]$
tan^{-1}x	$(-\infty, \infty)$	$\left(\frac{-\pi}{2}, \frac{\pi}{2}\right)$
csc^{-1}x	$(-\infty, \infty)$	$(0, \pi)$
sec^{-1}x	$(-\infty, -1) \cup [1, \infty)$	$\left[0, \frac{\pi}{2}\right) \cup \left(\frac{\pi}{2}, \pi\right]$
cot^{-1}x	$(-\infty, -1) \cup [1, \infty)$	$\left[\frac{-\pi}{2}, 0\right) \cup \left(0, \frac{\pi}{2}\right]$

TRIGONOMETRIC IDENTITIES

An **identity** is a relationship between two expressions that is *always true*. For example, $(x + a)^2 = x^2 + 2ax + a^2$ is an algebraic identity because regardless of the value of *x* or *a*, the expressions are always equivalent. In addition, if one were to graph both $y = (x + a)^2$ and $y = x^2 + 2ax + a^2$, the same graph would result. Trigonometric identities have these same qualities. There are many, many trigonometric identities, a couple dozen of which are important enough to commit to memory; these will be examined here.

RECIPROCAL IDENTITIES

$\csc\theta = \dfrac{1}{\sin\theta}$

$\sec\theta = \dfrac{1}{\cos\theta}$

$\cot\theta = \dfrac{1}{\tan\theta}$

QUOTIENT IDENTITIES

$\tan\theta = \dfrac{\sin\theta}{\cos\theta}$

$\cot\theta = \dfrac{\cos\theta}{\sin\theta}$

PYTHAGOREAN IDENTITIES

$\sin^2\theta + \cos^2\theta = 1$

$1 + \tan^2\theta = \sec^2\theta$

$1 + \cot^2\theta = \csc^2\theta$

CO–FUNCTION IDENTITIES

$\sin\left(\dfrac{\pi}{2} - \theta\right) = \cos\theta$

$\csc\left(\dfrac{\pi}{2} - \theta\right) = \sec\theta$

$\cos\left(\dfrac{\pi}{2} - \theta\right) = \sin\theta$

$\sec\left(\dfrac{\pi}{2} - \theta\right) = \csc\theta$

$\tan\left(\dfrac{\pi}{2} - \theta\right) = \cot\theta$

$\cot\left(\dfrac{\pi}{2} - \theta\right) = \tan\theta$

EVEN–ODD IDENTITIES

$\sin(u \pm v) = \sin(u)\cos(v) \pm \cos(u)\sin(v)$

$\cos(u \pm v) = \cos(u)\cos(v) \mp \sin(u)\sin(v)$

$\tan(u \pm v)\ \dfrac{\tan(u) \pm \tan(v)}{1 \mp (\tan(u)\tan(v))}$

DOUBLE–ANGLE IDENTITIES

$\cos 2\theta = \begin{cases} = \cos^2\theta - \sin^2\theta \\ = 2\cos^2\theta - 1 \\ = 1 - 2\sin^2\theta \end{cases}$

HALF–ANGLE IDENTITIES

$\sin A = \pm\sqrt{\dfrac{1 - \cos 2A}{2}}$

$\cos A = \pm\sqrt{\dfrac{1 + \cos 2A}{2}}$

As shown in the examples below, the trick in solving trigonometric identity problems is knowing which identity to use. There are no set rules that describe when to use each of the various identities, but there are a few general guidelines:

▶ Any time csc, sec, or cot are seen, use a reciprocal identity or a quotient identity to switch the expression to sin, cos, and tan.

▶ If an exponent is present, try the Pythagorean identities.

▶ A double angle should be replaced using the double-angle formulas.

SAMPLE QUESTIONS

78) If $\sin(u) = 0.4$ and $1 - \cos^2 u = z$, what is the value of z?

Answer:

$1 - \cos^2 u = z$ $\sin^2 u = z$ $0.4^2 = z$ **$z = 0.16$**	Rearrange this problem based on the identity $\sin^2 \theta + \cos^2 \theta = 1$.

79) If $\sin A = \frac{-3}{5}$ and $\cos B = \frac{12}{13}$, where $\pi \le A \le \frac{3\pi}{2}$ and $0 \le B \le \frac{\pi}{2}$, then find (a) $\sin(2A)$ and (b) $\cos(A + B)$.

Answer:

$\pi \le A \le \frac{3\pi}{2} \rightarrow$ sine and cosine are negative $\sin^2 A + \cos^2 A = 1$ $\left(-\frac{3}{5}\right)^2 + \cos^2 A = 1$ $\cos A = -\frac{4}{5}$ $0 \le B \le \frac{\pi}{2} \rightarrow$ sine and cosine are positive $\sin^2 B + \cos^2 B = 1$ $\sin^2 B + \left(\frac{12}{13}\right)^2 = 1$ $\sin B = \frac{5}{13}$	Find $\cos A$ and $\sin B$. Use the given domains to determine whether each value is positive or negative.
$\sin(2A) = 2 \sin A \cos A$ $= 2\left(-\frac{3}{5}\right)\left(-\frac{4}{5}\right)$ $= \frac{24}{25}$	Use the double angle identity to find $\sin(2A)$.
$\cos(A + B) = \cos A \cos B - \sin A \sin B$ $= \left(-\frac{4}{5}\right)\left(\frac{12}{13}\right) - \left(-\frac{3}{5}\right)\left(\frac{5}{13}\right)$ $= -\frac{33}{65}$	Use the sum identity to find $\cos(A + B)$.

80) Find the simplest form of: $\frac{\tan x - \sin x \cos x}{\tan x}$

Answer:

$\dfrac{\tan x}{\tan x} - \dfrac{\sin x \cos x}{\tan x}$	Break down the fraction in order to use identities. Split the fraction into two fractions with the same denominator.
$1 - \dfrac{\sin x \cos x}{\frac{\sin x}{\cos x}}$	Replace $\tan x$ using a quotient identity.
$1 - \dfrac{\sin x \cos x \cos x}{\sin x}$	Invert and multiply the denominator.
$1 - \cos^2 x$ $= \mathbf{\sin^2 x}$	$\sin x$ cancels, and then the Pythagorean identity can be used.

SOLVING TRIGONOMETRIC EQUATIONS

To solve trigonometric equations, isolate the trigonometric function using addition, subtraction, multiplication, division, or, in some cases, factoring. Sometimes it is necessary to use a trigonometric identity to simplify the expression.

Once the equation is down to a simple trig expression equaling a number, either use the special angles table in reverse, if one of the numbers in the table appears, or use the inverse trig feature on the calculator to find one solution. Often there will be more than one solution between 0 and 2π. The way to find the second solution is to find the other quadrant that would also have the correct sign for the trig function in question, and use a reference angle in that quadrant to figure out what the other angle value must be.

In general, because trigonometric functions are periodic, there will be infinitely many solutions that make the equation true. Often the question will only ask for solutions in a certain domain (usually $[0,2\pi]$), but if asked to find *all* solutions, simply find the solutions between 0 and 2π, and then add "+$2n\pi$" (radians) or "+$360n$" (degrees) for functions with a period of 2π, or + "$n\pi$" or "+$n180$" for functions with a domain of π. The n represents an integer and is therefore listing all the coterminal angles to the solution values, which would also be solutions.

SAMPLE QUESTION

81)   **Solve for θ. Give *all* solutions in degrees:** $\tan^2 \theta + \sec^2 \theta - 1 = 3.76$

Answer:

$\tan^2 \theta + \tan^2 \theta = 3.76$	
$2\tan^2 \theta = 3.76$ $\tan^2 \theta = 1.88$	Recognize that $\sec^2 \theta - 1$ is a rearrangement of a Pythagorean identity.
$\tan \theta = \pm 1.3711$ $\theta = \tan^{-1}(1.3711) = 53.9°$	Isolate the trig function.

| 53.9 degrees + n180 | Include in the answer all angles with a tangent of +1.3711 and -1.3711. |
| 126.1 degrees + n180 | |

NON-RIGHT TRIANGLES

Trigonometry is usually associated with right triangles. However, trigonometric functions can be used to solve for a missing side or angle in an oblique triangle (a triangle without any 90° angles). To solve an oblique triangle, either two sides and one angle, two angles and any one side, or all three sides must be known.

The Law of Sines is based on the proportionality between the sine of angles in a triangle and the sides of the triangle. It states, based on a triangle ABC with angles A, B, and C and corresponding sides a, b, c (a is the side opposite angle A, etc.):

$$\frac{a}{\sin A} = \frac{b}{\sin B} = \frac{c}{\sin C}$$

While all three proportionalities always hold, in practice only two of the three ratios are used at one time.

The Law of Cosines can be used when three sides are known, or when two sides and the included angle between those two sides are known. The law of cosines is based on Pythagorean identities:

$$c^2 = a^2 + b^2 - 2ab\cos(C)$$

$$b^2 = a^2 + c^2 - 2ac\cos(B)$$

$$a^2 = b^2 + c^2 - 2bc\cos(A)$$

SAMPLE QUESTIONS

82) **Find the degree measure of C in the following triangle, where $c = 12$, $b = 7$, and $a = 8$.**

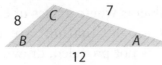

Answer:

Since three sides are given, use the law of cosines:

$12^2 = 8^2 + 7^2 - 2(8)(7)\cos C$	Begin with "$12^2 =$" because 12 is opposite the angle sought.
$144 = 64 + 49 - 112\cos C$	Simplify squares and products.
$31 = -112\cos C$	Subtract 64 and 49 from both sides.

$\cos C = \dfrac{31}{-112}$	Divide both sides by −112.
$= \cos^{-1}\left(-\dfrac{31}{112}\right)$ $= \mathbf{106.1°}$	Use inverse cosine to find the angle.

83) **If there is a triangle where $a = 110$, $c = 40$, and the measure of angle $A = 110°$, what is the measure of angle C?**

Answer:

Because an angle and the side opposite is known, and the side opposite the unknown angle is known, use the law of sines for this triangle:

$a = 110$ $A = 110°$ $c = 40$ $C = ?$	Identify the parts of the triangle given.
$\dfrac{40}{\sin C} = \dfrac{110}{\sin 110}$ $40(\sin 110) = 110\sin C$ $\dfrac{37.6}{110} = \sin C$ $.34 = \sin C$	Plug these values into the law of sines and solve for angle C.
$C = \sin^{-1}(.34)$ $= \mathbf{20°}$	Plug these values into the law of sines and solve for angle C (continued).

TRIGONOMETRIC WORD PROBLEMS AND MODELING WITH TRIGONOMETRIC FUNCTIONS

Trigonometry will often show up in word problems where angles of elevation or descent (or declination) are used. If a diagram is not provided, drawing one and carefully labeling it is advised. Trigonometric function models are used in situations where oscillatory, rotational, or other periodic behavior occurs. Constructing a graph of the situation usually helps in finding an equation that models it.

HELPFUL HINT

The pictures on the test may not be proportional, so don't rely on them when setting up equations.

SAMPLE QUESTIONS

84) **If a kite string is extended 10 feet and the kite is 8 feet in the air, what is the angle between the kite string and the ground?**

Answer:

The scenario described can be represented by this diagram:

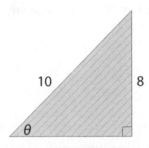

$\sin\theta = \dfrac{8}{10}$	Write an equation that relates angle θ to the opposite side and the hypotenuse.
$\theta = \sin^{-1}\dfrac{8}{10}$ $= \mathbf{53°}$	Use the inverse of sine to solve for the angle.

85) **The radius of a Ferris wheel is 20 feet, and the bottom of the wheel is 4 feet above the ground. When the chairs are all loaded, the first chair is the chair at the bottom of the wheel. The wheel begins turning clockwise and makes one complete revolution every 24 seconds (assume a constant rate). Construct a model (equation) that gives the height of the first chair, $h(t)$ as a function of time.**

Answer:

Begin solving this problem by creating a sketch of the wheel's motion for one rotation. Since at the beginning of the ride the first chair is the bottom chair, the graph should begin at 4. The peak of the graph will be 4 + 40 (add the diameter of the wheel) or 44. The graph will return to height 4 when t is 24 seconds.

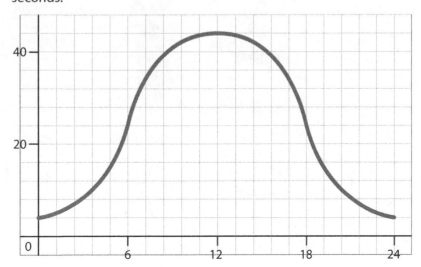

The midline can be found by averaging the maximum and minimum values, so $d = \frac{(4 + 44)}{2} = 24$.

The amplitude is the peak minus the midline value, so $a = 44 - 24 = 20$.

The period of the function is 24, so the b-value of the equation is $b = \frac{2\pi}{\text{period}} = \frac{2\pi}{24} = \frac{\pi}{12}$.

Since this function begins below the midline, it is a reflected cosine function. Putting all the pieces together, the height h of the function at any time can be modeled by:

$$h(t) = -20\cos(\tfrac{\pi}{12}t) + 24$$

CONIC SECTIONS

Conic sections refers to the two-dimensional shapes created when a plane intersects a cone. These intersections result in circles, ellipses, hyperbolas, or parabolas depending on the angle at which the plane crosses the cone.

Every conic section equation can be written using the general form for a conic section:

$$Ax^2 + Bxy + Cy^2 + Dx + Ey + F = 0$$

HELPFUL HINT

Only hyperbolas have a positive discriminant.

This equation contains two variables (x and y) that are both squared. The shape described by the equation can be determined by looking at its coefficients and the value of the conic section discriminant, which is defined as $B^2 - 4AC$.

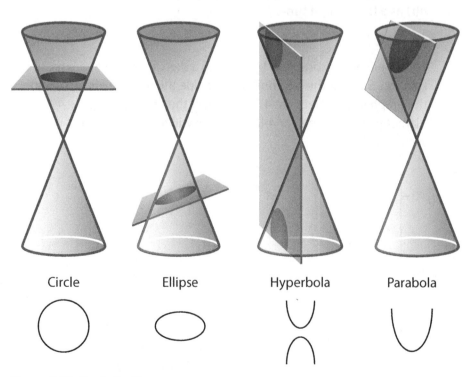

Circle   Ellipse   Hyperbola   Parabola

Figure 2.22. Conic Sections

Table 2.12. Identifying Conic Sections

$B^2 - 4AC < 0$ and $A = C$	circle
$B^2 - 4AC < 0$ and $A \neq C$	ellipse
$B^2 - 4AC > 0$	hyperbola
$B^2 - 4AC < 0$ and $A = 0$ or $C = 0$	parabola

SAMPLE QUESTION

86) Which conic section is described by the equation $x^2 + y^2 + 6x + 14y = 86$?

Answer:

$x^2 + y^2 + 6x + 14y = 86$	
$x^2 + y^2 + 6x + 14y - 86 = 0$	Put the equation in standard form.
$A = 1$ $B = 0$ $C = 1$	Identify the variables needed to solve for the discriminant.
$B^2 - 4AC =$ $0^2 - 4(1)(1) =$ -4	Solve for the discriminant.
$B^2 - 4AC < 0$ and $A = C$	**The equation describes a circle.**

CIRCLES

The standard form of a circle is written as $(x - h)^2 + (y - k)^2 = r^2$ where (h,k) is the center of the circle and r is its radius.

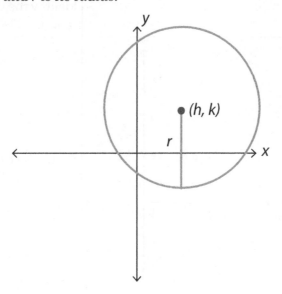

Figure 2.23. Graph of a Circle

SAMPLE QUESTIONS

87)   **What is the equation for the circle shown on the graph below?**

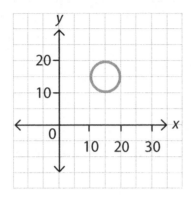

Answers:

$(h, k) = (15, 15)$ $r = 5$	Identify the center and radius of the circle.
$(x - h)^2 + (y - k)^2 = r^2$ $(x - 15)^2 + (y - 15)^2 = 25$	Plug these values into the standard equation for a circle.

88)   **Graph the equation:** $-10y + y^2 + x^2 - 38 = -18x$

Answer:

$-10y + y^2 + x^2 - 38 = -18x$ $x^2 + 18x + y^2 - 10y = 38$ $(x^2 + 18x + 81) + (y^2 - 10y + 25) = 38 + 106$ $(x + 9)^2 + (y - 5)^2 = 144$	Convert the equation to standard form by completing the square for each variable.
$(h, k) = (-9, 5)$ $r = 12$	Identify the circle's center and radius.

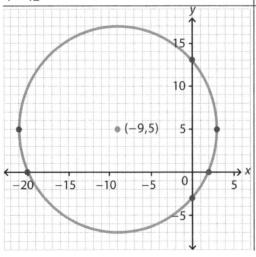

Graph the circle.

ELLIPSES

An **ellipse** is an oval-shaped conic section. A circle is a type of ellipse in the same way that a square is a type of rectangle: the similar key difference is that the proportions of an ellipse are not equal. An ellipse can either be wider than it is tall (a horizontal ellipse) or it can be taller than it is wide (a vertical ellipse).

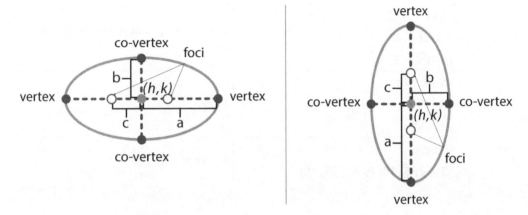

Figure 2.24. Horizontal and Vertical Ellipses

An ellipse has two fixed interior points known as **foci**. The sum of the distances between a focus and any point on the ellipse is constant. The foci always lie on the **major axis** of the ellipse, which is the longer axis. Each endpoint of the major axis is a **vertex**. The shorter axis is called the **minor axis** and its endpoints are the **co-vertices**.

The value a is the distance from the center to a vertex along the major axis, or one-half the major axis length. The distance b is from the center along the minor axis to a co-vertex, or one-half the minor axis length. Finally, the distance from the center of the ellipse to a focus is c. Together, these values can be used to solve for the center, foci, vertices, and co-vertices of an ellipse using the formulas in the table below.

> **HELPFUL HINT**
>
> a is always larger than b in an ellipse. When a is on the x-axis, the ellipse is horizontal; when a is on the y-axis, the ellipse is vertical.

Table 2.13. Formulas for Horizontal and Vertical Ellipses

Horizontal Ellipse	Vertical Ellipse
$\dfrac{(x-h)^2}{a^2} + \dfrac{(y-k)^2}{b^2} = 1$	$\dfrac{(x-h)^2}{b^2} + \dfrac{(y-k)^2}{a^2} = 1$
Center: (h,k)	Center: (h,k)
Foci: $(h \pm c, k)$	Foci: $(h, k \pm c)$
Vertices: $(h \pm a, k)$	Vertices: $(h, k \pm a)$
Co-Vertices: $(h, k \pm b)$	Co-Vertices: $(h \pm b, k)$

$$a > b$$
$$a^2 - b^2 = c^2$$

SAMPLE QUESTIONS

89) **What are the foci of an ellipse defined by the equation** $\frac{(x-5)^2}{64} + \frac{(y-10)^2}{9} = 1$?

Answer:

$(h, k) = (5, 10)$ $a = \sqrt{64} = 8$ $b = \sqrt{9} = 3$	Identify important values from the equation.
$a^2 - b^2 = c^2$ $64 - 9 = c^2$ $c = \sqrt{55}$	Find c.
$(h \pm c, k)$ $(5 + \sqrt{55}, 10)\ (5 - \sqrt{55}, 10)$	Find each focus.

90) **What is the equation of the ellipse graphed below.**

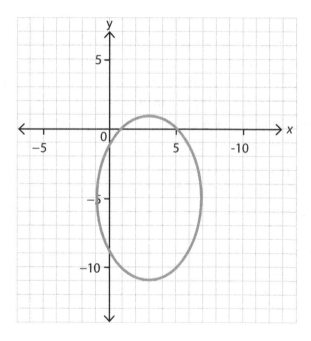

Answer:

center: $(3, -5)$ vertices: $(3, 1)$ and $(3, -11)$ co-vertices: $(-1, -5)$ and $(7, -5)$	Identify important points from the graph.

vertex:	
$(h, k \pm a)$	
$(3, -5 + a) = (3, 1)$	
$a = 6$	Use these points to find a and b. Alternatively, a and b can sometimes be found by counting on the graph.
co-vertex:	
$(h \pm b, k)$	
$(3 - b, -5) = (-1, -5)$	
$b = 4$	
$\dfrac{(x-3)^2}{16} + \dfrac{(y+5)^2}{36} = 1$	Plug h, k, a, and b into the standard equation for an ellipse.

HYPERBOLAS

A **hyperbola** is a pair of open curves—those that open toward the x-axis are horizontal hyperbolas, and those that open along the y-axis are vertical. Each curve of the hyperbola is called a branch.

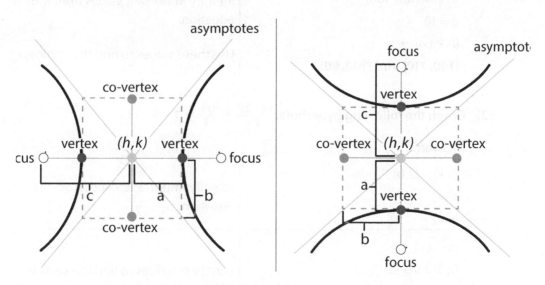

Figure 2.25. Horizontal and Vertical Hyperbolas

Hyperbolas have a center that lies halfway between the two curves, and two vertices, which are the points on each curve closest to the center. As in an ellipse, the distance from the center to the vertex is a, and the distance to the co-vertices is b. Unlike in an ellipse, however, a is not necessarily the larger value. The foci are located within each curve, and again the distance from the center to a focus is c. Hyperbolas also have a pair of **asymptotes** that cross at the center.

Table 2.14. Formulas for Horizontal and Vertical Hyperbolas

Horizontal Hyperbola	Vertical Hyperbola
$\dfrac{(x-h)^2}{a^2} - \dfrac{(y-k)^2}{b^2} = 1$	$\dfrac{(y-k)^2}{a^2} - \dfrac{(x-h)^2}{b^2} = 1$
Center: (h,k)	Center: (h,k)
Foci: $(h \pm c, k)$	Foci: $(h, k \pm c)$
Vertices: $(h \pm a, k)$	Vertices: $(h, k \pm a)$
Co-Vertices: $(h, k \pm b)$	Co-Vertices: $(h \pm b, k)$
Asymptotes: $y - k = \pm \frac{b}{a}(x-h)$	Asymptotes: $y - k = \pm \frac{a}{b}(x-h)$

$$a^2 + b^2 = c^2$$

SAMPLE QUESTIONS

91) **Find the vertices of the hyperbola graphed by the equation** $\dfrac{(y-100)^2}{100} - \dfrac{(x-100)^2}{100} = 1.$

Answer:

$(h,k) = (100, 100)$	Identify important values from the equation.
$a = 10$	
$(h, k \pm a) \rightarrow$	Use these values to find the vertices.
(100, 110) and (100, 90)	

92) **Graph the following hyperbola:** $\dfrac{(x-2)^2}{25} - \dfrac{(y)^2}{4} = 1$

Answer:

$(h,k) = (2,0)$	Identify important values from the equation.
$a = 5$	
$b = 2$	
$(h \pm a, k) \rightarrow$	Use these values to find the vertices and asymptotes.
$(7,0)$ and $(-3,0)$	
$y - k = \pm \frac{b}{a}(x-h) \rightarrow$	
$y = \frac{2}{5}(x-2)$ and $y = -\frac{2}{5}(x-2)$	

Use the vertices and asymptotes to graph the equation.

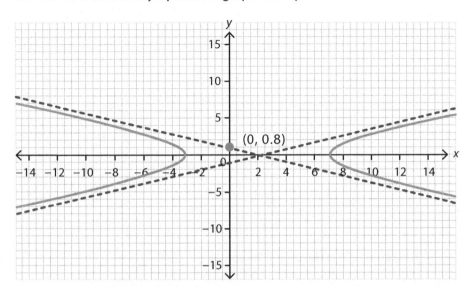

PARABOLAS

Parabolas are conic sections formed when only one variable in the general equation is squared. An equation in which the x is squared will open vertically (either up or down). Equations with the y variable squared result in a curve that opens horizontally (either right or left).

Like ellipses and hyperbolas, parabolas have a vertex and a focus, which is a point on the interior of the parabola. The distance from the vertex to the focus is a value called p. The **axis of symmetry** for the parabola will run through both the vertex and focus. Each parabola also has a directrix, which is a line that is a distance of p from the vertex in the opposite direction from the focus. The distance from the focus to any point on the parabola is the same as the distance from that point on the parabola to the directrix.

> **QUICK REVIEW**
>
> Why are vertical parabolas functions but horizontal ones are not?

Table 2.15. Formulas for Vertical and Horizontal Parabolas

Vertical Parabola	Horizontal Parabola
$(x - h)^2 = 4p(y - k)$ or	$(y - k)^2 = 4p(x - h)$ or
$y = a(x - h)^2 + k$	$x = a(y - k)^2 + h$
Vertex: (h, k)	Vertex: (h, k)
Focus: $(h, k + p)$ or $(h, k + \frac{1}{4a})$	Focus: $(h + p, k)$ or $(h + 1/4a, k)$
Axis of Symmetry: $x = h$	Axis of Symmetry: $y = k$
Directrix: $y = k - p$ or $y = k - \frac{1}{4a}$	Directrix: $x = h - p$ or $x = h - 1/4a$

$$p = \frac{1}{4a}$$

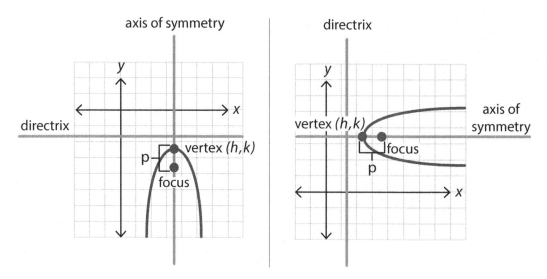

Figure 2.26. Vertical and Horizontal Parabolas

SAMPLE QUESTIONS

93) **What is the directrix of the parabola $y = -(x + 3)^2 - 2$?**

Answer:

$a = -1$ $k = -2$	Identify important values from the equation.
$y = k - \dfrac{1}{4a}$ $y = -2 - \dfrac{1}{4(-1)}$ $\boldsymbol{y = -1.75}$	Plug these values into the formula to find the directrix.

94) **What is the vertex of a parabola with a directrix of $x = -6$ and a focus of (2, 1)?**

Answer:

The vertex must be equidistant between focus and directrix and have the same *y* value as the focus, so the vertex is at (−2, 1).

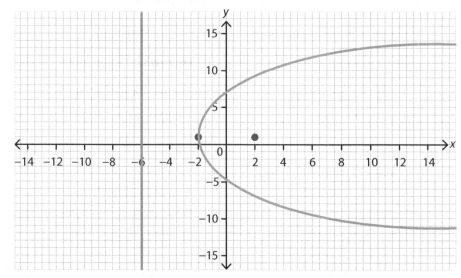

OTHER COORDINATE SYSTEMS

POLAR COORDINATE SYSTEM

Thus far, graphing and plotting have taken place in two-dimensions on the rectangular, or Cartesian, plane. The second most common way to graph or plot is by using polar coordinates. Graphing in **polar coordinates** is basically a "repackaging" of the ideas presented in the previous section on the unit circle, which involves an angle θ in standard position rotated within a circle with radius r. Each point in the polar plane can be described by the radius of the circle and by the angle, giving a coordinate pair (r, θ). The **radial coordinate** r gives the distance between the origin and the point, and the **angular coordinate** θ describes the angle of the point in standard position.

The point (r, θ) on the polar plane can be used to draw a right triangle with hypotenuse r and sides that correspond to the position of the x- and y-axis. Thus, points in the polar plane can be converted to the rectangular plane, and vice versa, using basic right triangle geometry.

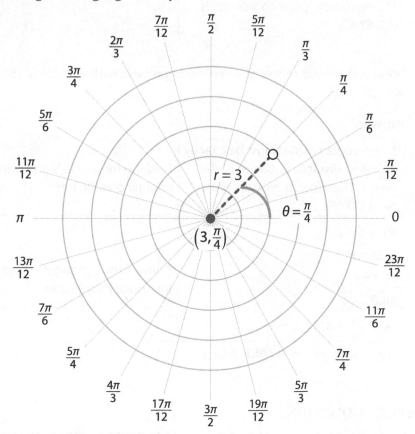

Figure 2.27. The Polar Plane

Table 2.16. Converting Between Rectangular and Polar Coordinates

To convert from (x, y) form to (r, θ)	To convert from (r, θ) form to (x, y)
$x^2 + y^2 = r^2$ $tan(\theta) = \frac{y}{x}$	$y = r\sin\theta$ $x = r\cos\theta$ $(x, y) = (r\cos\theta, \sin\theta)$

SAMPLE QUESTIONS

95) **Write the rectangular coordinate pair (5, 12) in polar form.**

Answer:

$x^2 + y^2 = r^2$ $5^2 + 12^2 = r^2$ $169 = r^2$ $13 = r$	Find r using the Pythagorean theorem.
$tan\,\theta = 12/5$ $tan^{-1}\left(\frac{12}{5}\right) = 67.4^o$ The equivalent polar coordinate is **(13, 67.4^o).**	Find θ using the definition of tangent.

96) **What is the polar center of a vertical hyperbola with vertices at (50, 100) and (50, 10)?**

Answer:

A hyperbola's center must be exactly the same distance between each vertex. Therefore, the vertex must be at (50, 55).	Find the hyperbola's center in rectangular coordinates.
$50^2 + 55^2 = r^2$ $5525 = r^2$ $74.3 = r$ $tan\theta = \frac{55}{50}$ $tan^{-1}\frac{55}{50} = 47.7^o$	Convert the rectangular coordinates into polar coordinates.
The polar center is at **(74.3, 47.7^o).**	

Parametric Equations

Another way to define functions and relations is using **parametric form**. This system defines every point (x, y) on a graph in terms of a third variable, called the **parameter** (usually t is used). Thus, the point is actually $(x(t), y(t))$. Often curves that are not functions (like ellipses and hyperbolas) are easier to define using parametric form. Also, by controlling which t values are used (the domain), parametrization allows for the graphing of only specific parts of curves.

HELPFUL HINT

There is a parametric option available on most graphing calculators, usually found in the "mode" section of the calculator.

Any function can parametrized by simply defining one variable in terms of the parameter, and then using substitution. For example, the function $y = 5x^2 + 3x$ can easily be made into a set of parametric equations that define the function by making $x(t) = t$ and $y(t) = 5t^2 + 3t$. To convert a parametric equation into rectangular form, it's necessary to "eliminate the parameter" by solving one of the parametric equations for t and replacing t in the other parametric equation with that quantity.

In physics, the t parameter often represents time. For instance, in projectile problems where an object is launched at a certain angle with some initial velocity, t represents time, $x(t)$ represents the object's horizontal position with respect to time, and $y(t)$ represents the object's vertical position with respect to time.

QUICK REVEW

What is the parameter used when points on the unit circle are defined as $x(\theta) = \cos\theta$ and $y(\theta) = \sin\theta$?

SAMPLE QUESTIONS

97) If a boulder falls from a cliff, how many seconds will it take for it to hit the ground if $y(t) = 1000 - 9.8\left(\frac{t^2}{2}\right)$ where $y(t)$ is the boulder's vertical position?

Answer:

$0 = 1000 - 9.8\left(\frac{t^2}{2}\right)$

$\frac{-1000}{-9.8} = \frac{t^2}{2}$

$102 = \frac{t^2}{2}$

$t^2 = 204$

$t = 14.3$

When the boulder hits the ground, the vertical component will be 0, so set $y(t) = 0$.

The boulder will hit the ground in 14.3 seconds.

98) What are the rectangular coordinates of the roots of the parametric equations $x(t) = \sqrt{t}$ and $y(t) = t - 6$?

Answer:

$x(t) = \sqrt{t}$

$x = \sqrt{t}$

$t = x^2$

Solve one equation for t.

$y(t) = t - 6$ $y = t - 6$ $y = x^2 - 6$	Substitute this value into the second equation.
$0 = x^2 - 6$ $\boldsymbol{x = \pm\sqrt{6}}$	Find the rectangular equation's roots.

3

Geometry

Geometry is the study of shapes, angles, volumes, areas, lines, points, and the relationships among them. It is normally approached as an axiomatic system; that is, a small number of entities are taken for granted as true, and everything else is derived logically from them.

EQUALITY, CONGRUENCE, AND SIMILARITY

When discussing shapes in geometry, the term **congruent** is used to mean that two shapes have the same shape and size (but not necessarily the same orientation or location). This concept is slightly different from equality, which is used in geometry to describe numerical values. For example, if the length of two lines are equal, the two lines themselves are called congruent. Congruence is written using the symbol ≅. On figures, congruent parts are denoted with hash marks.

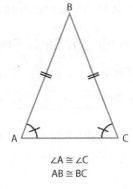

∠A ≅ ∠C
AB ≅ BC

Figure 3.1. Congruent Parts of a Triangle

Shapes which are **similar** have the same shape but the not the same size, meaning their corresponding angles are the same but their lengths are not. For two shapes to be similar, the ratio of their corresponding sides must be a constant (usually written as k). Similarity is described using the symbol ~.

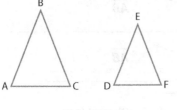

ABC ~ DEF

$$\frac{AB}{DE} = \frac{BC}{EF} = \frac{AC}{DF}$$

Figure 3.2. Similar Triangles

PROPERTIES OF SHAPES

BASIC DEFINITIONS

The basic figures from which many other geometric shapes are built are points, lines, and planes. A **point** is a location in a plane. It has no size or shape, but is represented by a dot. It is labeled using a capital letter.

A **line** is a one-dimensional collection of points that extends infinitely in both directions. At least two points are needed to define a line, and any points that lie on the same line are **colinear**. Lines are represented by two points, such as *A* and *B*, and the line symbol: ($\overleftrightarrow{AB}$). Two lines on the same plane will intersect unless they are **parallel**, meaning they have the same slope. Lines that intersect at a 90 degree angle are **perpendicular**.

A **line segment** has two endpoints and a finite length. The length of a segment, called the measure of the segment, is the distance from *A* to *B*. A line segment is a subset of a line, and is also denoted with two points, but with a segment symbol: ($\overline{AB}$). The **midpoint** of a line segment is the point at which the segment is divided into two equal parts. A line, segment, or plane that passes through the midpoint of a segment is called a **bisector** of the segment, since it cuts the segment into two equal segments.

A **ray** has one endpoint and extends indefinitely in one direction. It is defined by its endpoint, followed by any other point on the ray: $\overrightarrow{AB}$. It is important that the first letter represents the endpoint. A ray is sometimes called a half line.

Table 3.1. Basic Geometric Figures

Term	Dimensions	Graphic	Symbol
point	zero	●	$\cdot A$
line segment	one	A ———— B	$\overline{AB}$
ray	one	A ———→ B	$\overrightarrow{AB}$
line	one	←————→	$\overleftrightarrow{AB}$
plane	two	▱	Plane *M*

A **plane** is a flat sheet that extends indefinitely in two directions (like an infinite sheet of paper). A plane is a two-dimensional (2D) figure. A plane can always be defined through any three noncollinear points in three-dimensional (3D) space. A plane is named using any three points that are in the plane (for example, plane

ABC). Any points lying in the same plane are said to be **coplanar**. When two planes intersect, the intersection is a line.

1) Which points and lines are not contained in plane *M* in the diagram below?

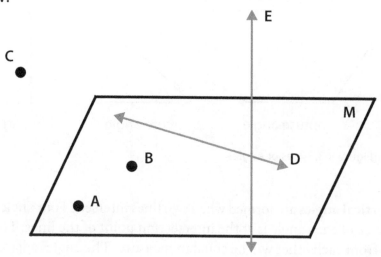

Answer:

Points *A* and *B* and line *D* are all on plane *M*. Point *C* is above the plane, and line *E* cuts through the plane and thus does not lie on plane *M*. The point at which line *E* intersects plane *M* is on plane *M* but the line as a whole is not.

ANGLES

Angles are formed when two rays share a common endpoint. They are named using three letters, with the vertex point in the middle (for example ∠*ABC*, where *B* is the vertex). They can also be labeled with a number or named by their vertex alone (if it is clear to do so). Angles are also classified based on their angle measure. A **right angle** has a measure of exactly 90°. **Acute angles** have measures that are less than 90°, and **obtuse angles** have measures that are greater than 90°.

HELPFUL HINT
Angles can be measured in degrees or radian. Use the conversion factor 1 rad = 57.3 degrees to convert between them.

Any two angles that add to make 90° are called **complementary angles**. A 30° angle would be complementary to a 60° angle. **Supplementary angles** add up to 180°. A supplementary angle to a 60° angle would be a 120° angle; likewise, 60° is the **supplement** of 120°. The complement and supplement of any angle must always be positive. For example, a 140 degree has no complement. Angles that are next to each other and share a common ray are called **adjacent angles**. Angles that are adjacent and supplementary are called a **linear pair** of angles. Their nonshared rays form a line (thus the *linear* pair). Note that angles that are supplementary do not need to be adjacent; their measures simply need to add to 180°.

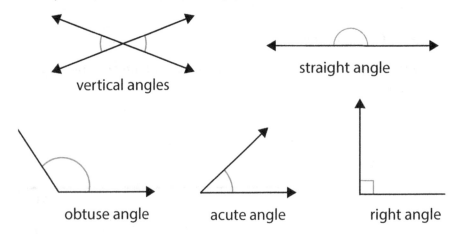

Figure 3.3. Types of Angles

Vertical angles are formed when two lines intersect. Four angles will be formed; the vertex of each angle is at the intersection point of the lines. The vertical angles across from each other will be equal in measure. The angles adjacent to each other will be linear pairs and therefore supplementary.

A ray, line, or segment that divides an angle into two equal angles is called an **angle bisector**.

SAMPLE QUESTIONS

2) How many linear pairs of angles are there in the following figure?

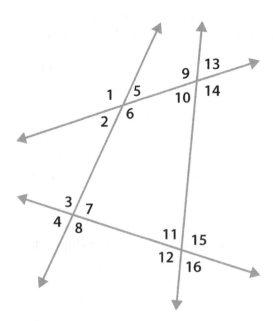

Answers:

Any two adjacent angles that are supplementary are linear pairs, so there are 16 linear pairs in the figure ($\angle 1$ and $\angle 5$, $\angle 2$ and $\angle 6$, $\angle 5$ and $\angle 6$, $\angle 2$ and $\angle 1$, and so on).

3)   If angles M and N are supplementary and $\angle M$ is 30° less than twice $\angle N$, what is the degree measurement of each angle?

Answer:

$\angle M + \angle N = 180°$ $\angle M = 2\angle N - 30°$	Set up a system of equations.
$\angle M + \angle N = 180°$ $(2\angle N - 30°) + \angle N = 180°$ $3\angle N - 30° = 180°$ $3\angle N = 210°$ $\angle N = 70°$	Use substitution to solve for $\angle N$.
$\angle M + \angle N = 180°$ $\angle M + 70° = 180°$ $\boldsymbol{\angle M = 110°}$	Solve for $\angle M$ using the original equation.

CIRCLES

A **circle** is the set of all the points in a plane that are the same distance from a fixed point called the **center**. The distance from the center to any point on the circle is the **radius** of the circle. The distance around the circle (the perimeter) is called the **circumference**.

The ratio of a circle's circumference to its diameter is a constant value called pi (π), an irrational number which is commonly rounded to 3.14. The formula to find a circle's circumference is $C = 2\pi r$. The formula to find the enclosed area of a circle is $A = \pi r^2$.

Circles have a number of unique parts and properties:

▸   The **diameter** is the largest measurement across a circle. It passes through the circle's center, extending from one side of the circle to the other. The measure of the diameter is twice the measure of the radius.

▸   A line that cuts across a circle and touches it twice is called a **secant** line. The part of a secant line that lies within a circle is called a **chord**.

Two chords within a circle are of equal length if they are are the same distance from the center.

▶ A line that touches a circle or any curve at one point is **tangent** to the circle or the curve. These lines are always exterior to the circle. A line tangent to a circle and a radius drawn to the point of tangency meet at a right angle (90°).

▶ An **arc** is any portion of a circle between two points on the circle. The **measure** of an arc is in degrees, whereas the **length of the arc** will be in linear measurement (such as centimeters or inches). A **minor arc** is the small arc between the two points (it measures less than 180°), whereas a **major arc** is the large arc between the two points (it measures greater than 180°).

▶ An angle with its vertex at the center of a circle is called a **central angle**. For a central angle, the measure of the arc intercepted by the sides of the angle (in degrees) is the same as the measure of the angle.

▶ A **sector** is the part of a circle *and* its interior that is inside the rays of a central angle (its shape is like a slice of pie).

	Area of Sector	Length of an Arc
Degrees	$A = \frac{\theta}{360°} \times \pi r^2$	$s = \frac{\theta}{360°} \times 2\pi r$
Radians	$A = \frac{1}{2}\pi^2\theta$	$s = r\theta$

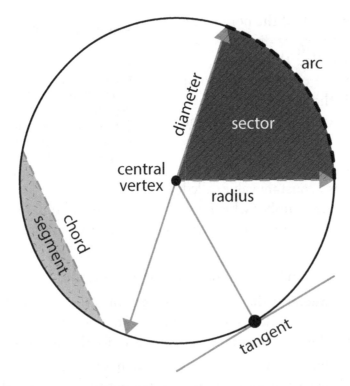

Figure 3.4. Parts of a Circle

▶ An **inscribed angle** has a vertex on the circle and is formed by two chords that share that vertex point. The angle measure of an inscribed angle is one-half the angle measure of the central angle with the same endpoints on the circle.

▶ A **circumscribed angle** has rays tangent to the circle. The angle lies outside of the circle.

▶ Any angle outside the circle, whether formed by two tangent lines, two secant lines, or a tangent line and a secant line, is equal to half the difference of the intercepted arcs.

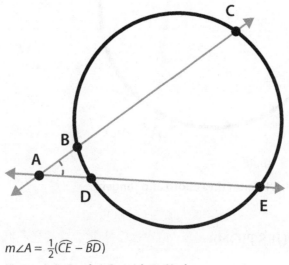

$$m\angle A = \tfrac{1}{2}(\overset{\frown}{CE} - \overset{\frown}{BD})$$

Figure 3.5. Angles Outside a Circle

▶ Angles are formed within a circle when two chords intersect in the circle. The measure of the smaller angle formed is half the sum of the two smaller arc measures (in degrees). Likewise, the larger angle is half the sum of the two larger arc measures.

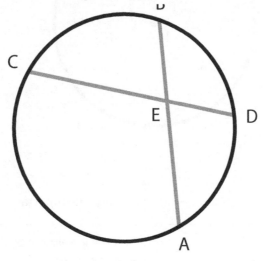

$$m\angle E = \tfrac{1}{2}(\overset{\frown}{AC} + \overset{\frown}{BD})$$

Figure 3.6. Intersecting Chords

▶ If a chord intersects a line tangent to the circle, the angle formed by this intersection measures one half the measurement of the intercepted arc (in degrees).

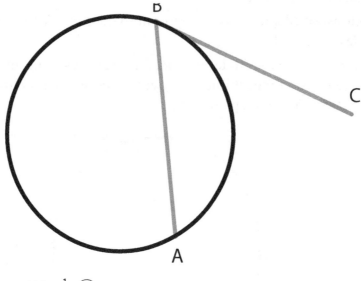

$m\angle ABC = \frac{1}{2}m\widehat{AB}$

Figure 3.7. Intersecting Chord and Tangent

SAMPLE QUESTIONS

4) Find the area of the sector *NHS* of the circle below with center at *H*:

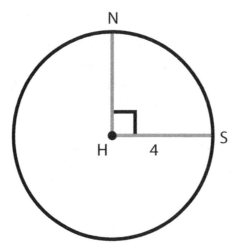

Answer:

$r = 4$ $\angle NHS = 90°$	Identify the important parts of the circle.
$A = \frac{\theta}{360°} \times \pi r^2$ $= \frac{90}{360} \times \pi(4)^2$	Plug these values into the formula for the area of a sector.

$$= \frac{1}{4} \times 16\pi$$

$$= 4\pi$$

> Plug these values into the formula for the area of a sector (continued).

5) In the circle below with center O, the minor arc ACB measures 5 feet. What is the measurement of $m\angle AOB$?

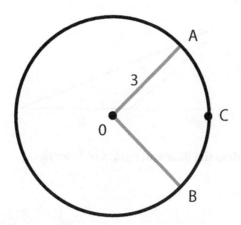

Answer:

$r = 3$ length of $\overline{ACB} = 5$	Identify the important parts of the circle.
$s = \frac{\theta}{360°} \times 2\pi r$ $5 = \frac{\theta}{360} \times 2\pi(3)$ $\frac{5}{6\pi} = \frac{\theta}{360}$ $\theta = 95.5°$ $m\angle AOB = 95.5°$	Plug these values into the formula for the length of an arc and solve for θ.

TRIANGLES

Much of geometry is concerned with triangles as they are commonly used shapes. A good understanding of triangles allows decomposition of other shapes (specifically polygons) into triangles for study.

Triangles have three sides, and the three interior angles always sum to 180°. The formula for the area of a triangle is $A = \frac{1}{2}bh$ or one-half the product of the base and height (or altitude) of the triangle.

Some important segments in a triangle include the angle bisector, the altitude, and the median. The **angle bisector** extends from the side opposite an angle to bisect that angle. The **altitude** is the shortest distance from a vertex of the triangle to the line containing the base side opposite that vertex. It is perpendicular to that line and can occur on the outside of the triangle. The **median** extends from an angle to bisect the opposite side.

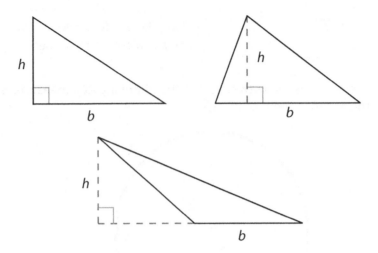

Figure 3.8. Finding the Base and Height of Triangles

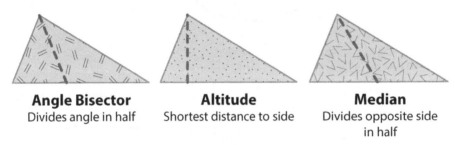

Angle Bisector
Divides angle in half

Altitude
Shortest distance to side

Median
Divides opposite side
in half

Figure 3.9. Important Segments in a Triangle

Triangles have two "centers." The **orthocenter** is formed by the intersection of a triangle's three altitudes. The **centroid** is where a triangle's three medians meet.

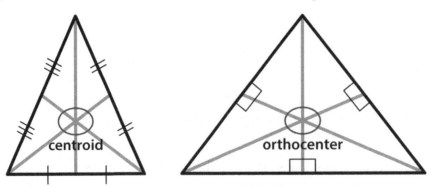

Figure 3.10. Centroid and Orthocenter of a Triangle

Triangles can be classified in two ways: by sides and by angles.

A **scalene triangle** has no equal sides or angles. An **isosceles triangle** has two equal sides and two equal angles, often called **base angles**. In an **equilateral triangle**, all three sides are equal as are all three angles. Moreover, because the sum of the angles of a triangle is always 180°, each angle of an equilateral triangle must be 60°.

Triangles Based on Sides

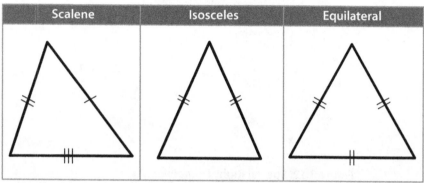

Triangles Based on Angles

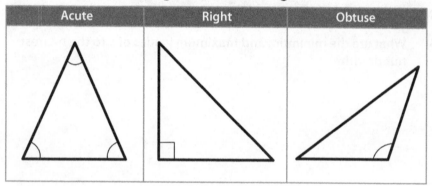

Figure 3.11. Types of Triangles

A **right triangle** has one right angle (90°) and two acute angles. An **acute triangle** has three acute angles (all angles are less than 90°). An **obtuse triangle** has one obtuse angle (more than 90°) and two acute angles.

For any triangle, the side opposite the largest angle will have the longest length, while the side opposite the smallest angle will have the shortest length. The **triangle inequality theorem** states that the sum of any two sides of a triangle must be greater than the third side. If this inequality does not hold, then a triangle cannot be formed. A consequence of this theorem is the **third-side rule**: if b and c are two sides of a triangle, then the measure of the third side a must be between the sum of the other two sides and the difference of the other two sides: $c - b < a < c + b$.

HELPFUL HINT

Trigonometric functions can be employed to find missing sides and angles of a triangle.

Solving for missing angles or sides of a triangle is a common type of triangle problem. Often a right triangle will come up on its own or within another triangle. The relationship among a right triangle's sides is known as the **Pythagorean theorem**: $a^2 + b^2 = c^2$, where c is the hypotenuse and is across from the 90° angle.

Right triangles with angle measurements of 90° – 45° – 45° and 90° – 60° – 30° are known as "special" right triangles and have specific relationships between their sides and angles.

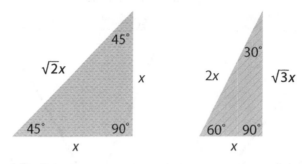

Figure 3.12. Special Right Triangles

SAMPLE QUESTIONS

6)   **What are the minimum and maximum values of *x* to the nearest hundredth?**

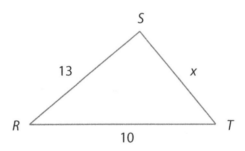

Answers:

The sum of two sides is 23 and their difference is 3. To connect the two other sides and enclose a space, *x* must be less than the sum and greater than the difference (that is, $3 < x < 23$). Therefore, ***x*'s minimum value to the nearest hundredth is 3.01 and its maximum value is 22.99.**

7)   **Examine and classify each of the following triangles:**

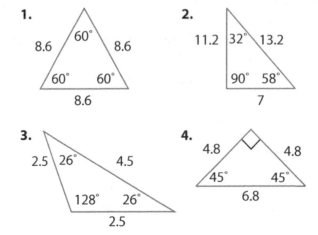

Answers:

Triangle 1 is an equilateral triangle (all 3 sides are equal, and all 3 angles are equal)

Triangle 2 is a scalene, right triangle (all 3 sides are different, and there is a 90° angle)

Triangle 3 is an obtuse, isosceles triangle (there are 2 equal sides and, consequently, 2 equal angles)

Triangle 4 is a right, isosceles triangle (there are 2 equal sides and a 90° angle)

8) **Given the diagram, if** $XZ = 100$, $WZ = 80$, **and** $XU = 70$, **then** $WY = ?$

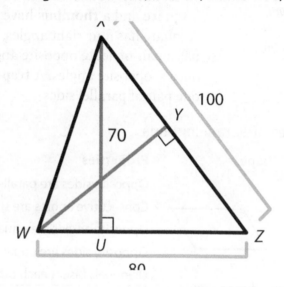

Answer:

$WZ = b_1 = 80$

$XU = h_1 = 70$

$XZ = b_2 = 100$

$WY = h_2 = ?$

$A = \frac{1}{2} bh$

$A_1 = \frac{1}{2}(80)(70) = 2800$

$A_2 = \frac{1}{2}(100)(h_2)$

The given values can be used to write two equation for the area of $\triangle WXZ$ with two sets of bases and heights.

$2800 = \frac{1}{2}(100)(h_2)$

$h_2 = 56$

$WY = 56$

Set the two equations equal to each other and solve for WY.

Quadrilaterals

All closed, four-sided shapes are **quadrilaterals**. The sum of all internal angles in a quadrilateral is always 360°. (Think of drawing a diagonal to create two triangles. Since each triangle contains 180°, two triangles, and therefore the quadrilateral, must contain 360°.) The **area of any quadrilateral** is $A = bh$, where b is the base and h is the height (or altitude).

A **parallelogram** is a quadrilateral with two pairs of parallel sides. A rectangle is a parallelogram with two pairs of equal sides and four right angles. A **kite** also has two pairs of equal sides, but its equal sides are consecutive. Both a **square** and a **rhombus** have four equal sides. A square has four right angles, while a rhombus has a pair of acute opposite angles and a pair of obtuse opposite angles. A **trapezoid** has exactly one pair of parallel sides.

> **HELPFUL HINT**
>
> All squares are rectangles and all rectangles are parallelograms; however, not all parallelograms are rectangles and not all rectangles are squares.

Table 3.2 Properties of Parallelograms

Term	Shape	Properties
Parallelogram		Opposite sides are parallel.
		Consecutive angles are supplementary.
		Opposite angles are equal.
		Opposite sides are equal.
		Diagonals bisect each other.
Rectangle		All parallelogram properties hold.
		Diagonals are congruent *and* bisect each other.
		All angles are right angles.
Square		All rectangle properties hold.
		All four sides are equal.
		Diagonals bisect angles.
		Diagonals intersect at right angles and bisect each other.
Kite		One pair of opposite angles is equal.
		Two pairs of consecutive sides are equal.
		Diagonals meet at right angles.
Rhombus		All four sides are equal.
		Diagonals bisect angles.
		Diagonals intersect at right angles and bisect each other.

Term	Shape	Properties
Trapezoid		One pair of sides is parallel. Bases have different lengths. Isosceles trapezoids have a pair of equal sides (and base angles).

SAMPLE QUESTIONS

9) **In parallelogram *ABCD*, the measure of angle *m* is is *m*° = 260°. What is the measure of *n*°?**

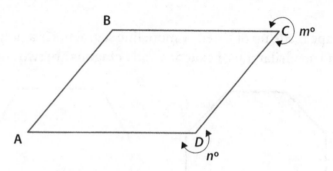

Answers:

$260° + m\angle C = 360°$ $m\angle C = 100°$	Find $\angle C$ using the fact that the sum of $\angle C$ and m is 360°.
$m\angle C + m\angle D = 180°$ $100° + m\angle D = 180°$ $m\angle D = 80°$	Solve for $\angle D$ using the fact that consecutive interior angles in a quadrilateral are supplementary.
$m\angle D + n = 360°$ **$n = 280°$**	Solve for *n* by subtracting $m\angle D$ from 360°.

10) **A rectangular section of a football field has dimensions of *x* and *y* and an area of 1000 square feet. Three additional lines drawn vertically divide the section into four smaller rectangular areas as seen in the diagram below. If all the lines shown need to be painted, calculate the total number of linear feet, in terms of *x*, to be painted.**

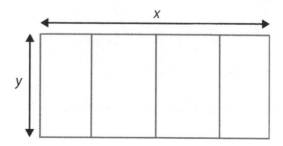

Answer:

$A = 1000 = xy$ $L = 2x + 5y$	Find equations for the area of the field and length of the lines to be painted (L) in terms of x and y.
$y = \frac{1000}{x}$ $L = 2x + 5y$ $L = 2x + 5\left(\frac{1000}{x}\right)$ $\mathbf{L = 2x + \frac{5000}{x}}$	Substitute to find L in terms of x.

Polygons

Any closed shape made up of three or more line segments is a polygon. In addition to triangles and quadrilaterals, **hexagons** and **octagons** are two common polygons.

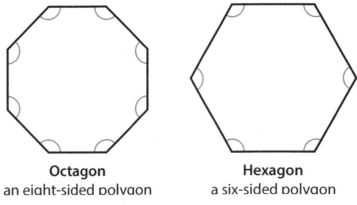

Octagon
an eight-sided polygon

Hexagon
a six-sided polygon

Figure 3.13. Common Polygons

The two polygons depicted above are **regular polygons**, meaning that they are equilateral (all sides having equal lengths) and equiangular (all angles having equal measurements). Angles inside a polygon are **interior angles**, whereas those formed by one side of the polygon and a line extending outside the polygon are **exterior angles**:

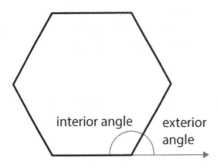

interior angle   exterior angle

Figure 3.14 Interior and Exterior Angles

The sum of the all the exterior angles of a polygon is always 360°. Dividing 360° by the number of a polygon's sides finds the measure of the polygon's exterior angles.

To determine the sum of a polygon's interior angles, choose one vertex and draw diagonals from that vertex to each of the other vertices, decomposing the polygon into multiple triangles. For example, an octagon has six triangles within it, and therefore the sum of the interior angles is 6 × 180° = 1080°. In general, the formula for finding the sum of the angles in a polygon is *sum of angles* = (n – 2) × 180°, where *n* is the number of sides of the polygon.

To find the measure of a single interior angle in a regular polygon, simply divide the sum of the interior angles by the number of angles (which is the same as the number of sides). So, in the octagon example, each angle is $\frac{1080}{8}$ = 135°.

In general, the formula to find the measure of a regular polygon's interior angles is: *interior angle* $= \frac{(n-2)}{n} \times 180°$ where *n* is the number of sides of the polygon.

To find the area of a polygon, it is helpful to know the perimeter of the polygon (*p*), and the **apothem** (*a*). The apothem is the shortest (perpendicular) distance from the polygon's center to one of the sides of the polygon. The formula for the area is: *area* $= \frac{ap}{2}$.

Finally, there is no universal way to find the perimeter of a polygon (when the side length is not given). Often, breaking the polygon down into triangles and adding the base of each triangle all the way around the polygon is the easiest way to calculate the perimeter.

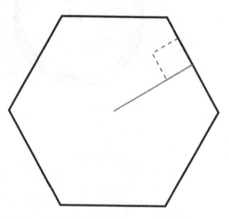

Figure 3.15. Apothem in a Hexagon

11) **What is the measure of an exterior angle and an interior angle of a regular 400-gon?**

Answer:

The sum of the exterior angles is 360°. Dividing this sum by 400 gives $\frac{360°}{400}$ = **0.9°**. Since an interior angle is supplementary to an exterior angle, all the interior angles have measure 180 – 0.9 = **179.1°**. Alternately, using the formula for calculating the interior angle gives the same result:

interior angle $= \frac{400-2}{400} \times 180° = 179.1°$

12) The circle and hexagon below both share center point T. The hexagon is entirely inscribed in the circle. The circle's radius is 5. What is the area of the shaded area?

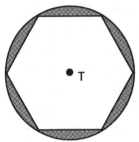

Answer:

$A_c = \pi r^2$ $= \pi(5)^2$ $= 25\pi$	The area of the shaded region will be the area of the circle minus the area of the hexagon. Use the radius to find the area of the circle.
$a = 2.5\sqrt{3}$ $A_H = \dfrac{ap}{2}$ $= \dfrac{(2.5\sqrt{3})(30)}{2}$ $= 64.95$	To find the area of the hexagon, draw a right triangle from the vertex, and use special right triangles to find the hexagon's apothem. Then, use the apothem to calculate the area.
$= A_c - A_H$ $= 25\pi - 2.5\sqrt{3}$ $\approx \mathbf{13.59}$	Subtract the area of the hexagon from the circle to find the area of the shaded region.

THREE–DIMENSIONAL SOLIDS

PROPERTIES OF THREE–DIMENSIONAL SOLIDS

Three-dimensional solids have depth in addition to width and length. **Volume** is expressed as the number of cubic units any solid can hold—that is, what it takes to fill it up. **Surface area** is the sum of the areas of the two-dimensional figures that are found on its surface. Some three-dimensional shapes also have a unique property called a slant height (l), which is the distance from the base to the apex along a lateral face.

Table 3.3 Three–Dimensional Shapes and Formulas

Term	Shape	Formula	
Prism		$V = Bh$ $SA = 2lw + 2wh + 2lh$ $d^2 = a^2 + b^2 + c^2$	B = area of base h = height l = length w = width d = longest diagonal
Cube		$V = s^3$ $SA = 6s^2$	s = cube edge
Sphere		$V = \frac{4}{3}\pi r^3$ $SA = 4\pi r^2$	r = radius
Cylinder		$V = Bh = \pi r^2 h$ $SA = 2\pi r^2 + 2\pi rh$	B = area of base h = height r = radius
Cone		$V = \frac{1}{3}\pi r^2 h$ $SA = \pi r^2 + \pi rl$	r = radius h = height l = slant height
Pyramid		$V = \frac{1}{3}Bh$ $SA = B + \frac{1}{2}(p)l$	B = area of base h = height p = perimeter l = slant height

Finding the surface area of a three-dimensional solid can be made easier by using a **net**. This two-dimensional "flattened" version of a three-dimensional shape shows the component parts that comprise the surface of the solid.

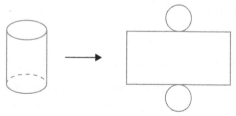

Figure 3.16. Net of a Cylinder

SAMPLE QUESTIONS

13) A sphere has a radius *z*. If that radius is increased by *t*, by how much is the surface area increased? Write the answer in terms of *z* and *t*.

Answer:

$SA_1 = 4\pi z^2$	Write the equation for the area of the original sphere.
$SA_2 = 4\pi(z + t)^2$ $= 4\pi(z^2 + 2zt + t^2)$ $= 4\pi z^2 + 8\pi zt + 4\pi t^2$	Write the equation for the area of the new sphere.
$A_2 - A_1 = 4\pi z^2 + 8\pi zt + 4\pi t^2 - 4\pi z^2$ $= \mathbf{4\pi t^2 + 8\pi zt}$	To find the difference between the two, subtract the original from the increased surface area:

14) A cube with volume 27 cubic meters is inscribed within a sphere such that all of the cube's vertices touch the sphere. What is the length of the sphere's radius?

Answer:

Since the cube's volume is 27, each side length is equal to $\sqrt[3]{27} = 3$. The long diagonal distance from one of the cube's vertices to its opposite vertex will provide the sphere's diameter:

$$d = \sqrt{3^2 + 3^2 + 3^2} = \sqrt{27} = 5.2$$

Half of this length is the radius, which is **2.6 meters**.

CONGRUENCE AND SIMILARITY IN THREE-DIMENSIONAL SHAPES

Three-dimensional shapes may also be congruent if they are the same size and shape, or similar if their corresponding parts are proportional. For example, a pair of cones is similar if the ratios of the cones' radii and heights are proportional. For rectangular prisms, all three dimensions must be proportional for the prisms to be similar. If two shapes are similar, their corresponding areas and volumes will also be proportional. If the constant of proportionality of the linear measurements of a 3D shape is k, the constant of proportionality between the areas will be k^2, and the constant of proportionality between the volumes will be k^3.

All spheres are similar as a dilation of the radius of a sphere will make it equivalent to any other sphere.

SAMPLE QUESTIONS

15) A square-based pyramid has a height of 10 cm. If the length of the side of the square is 6 cm, what is the surface area of the pyramid?

Answer:

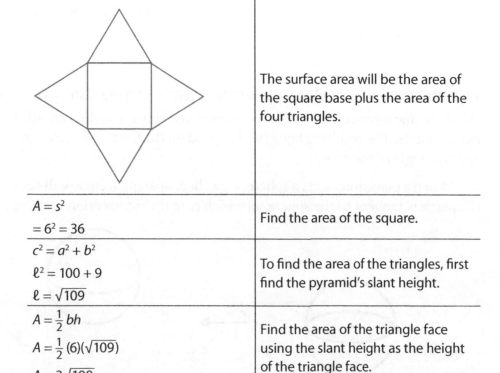

	The surface area will be the area of the square base plus the area of the four triangles.
$A = s^2$ $= 6^2 = 36$	Find the area of the square.
$c^2 = a^2 + b^2$ $\ell^2 = 100 + 9$ $\ell = \sqrt{109}$	To find the area of the triangles, first find the pyramid's slant height.
$A = \frac{1}{2} bh$ $A = \frac{1}{2}(6)(\sqrt{109})$ $A = 3\sqrt{109}$	Find the area of the triangle face using the slant height as the height of the triangle face.
$SA = 36 + 4(3\sqrt{109})$ $\approx$ **161.3 cm²**	Add the area of the square base and the four triangles to find the total surface area.

16) Given that two cones are similar and one cone's radius is three times longer than the other's radius, what is the volume of the smaller cone if the larger cone has a volume of 81 π cubic inches and a height of 3 inches?

Answer:

$V_1 = 81\pi$ $h_1 = 3$	Identify the given variables.
$V_1 = \frac{1}{3}\pi r_1^2 h_1$ $81\pi = \frac{1}{3}\pi(r_1)(3)$ $r_1 = 9$	Find the radius of the larger cone with the given information.
$r_2 = \frac{1}{3} r_1$ $r_2 = \frac{1}{3}(9)$ $r_2 = 3$ $h_2 = \frac{1}{3} h_1$ $h_2 = \frac{1}{3}(3)$ $h_2 = 1$	Use the given scale factor to find the second cone's radius and height.

$V_2 = \frac{1}{3}\pi r_2^2 h_2$

$V_2 = \frac{1}{3}\pi(3)^2(1)$

$\boldsymbol{V_2 = 3\pi}$

Find the area of the smaller cone.

INTERSECTIONS OF PLANES WITH THREE-DIMENSIONAL FIGURES

When a plane intersects with a three-dimensional figure, usually a two-dimensional figure results. The resulting figure will depend on the the three-dimensional shape and the angle of the plane.

When a plane intersects a sphere, regardless of angle, a circle will result, unless the plane is tangent to the sphere, in which case the intersection is a point.

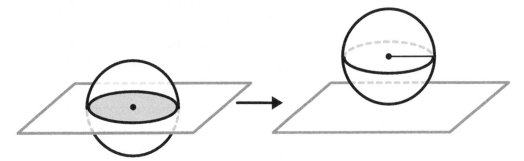

Figure 3.17. Plane Intersecting a Circle

When a plane intersects a cylinder, several different intersections may occur:

▶ A circle will result if the plane is parallel to the base of the cylinder.

▶ A line will result if the plane is tangent to the side of the cylinder.

▶ A rectangle will result if the plane is perpendicular to the cylinder.

▶ An ellipse will result if the plane intersects at an angle to the side of the cylinder.

A similar array of possibilities occurs when a plane intersects a rectangular prism:

▶ A rectangle will occur if planes are intersecting parallel to a side.

▶ A square may occur if the plane is taken at an angle through an edge such that the cut length is the same as the edge length.

▶ A triangle may occur if a corner of the prism is cut off by the intersecting plane.

▶ A point may be the intersection if the plane is at an angle and only intersects a single point of a corner.

▶ A line may occur if the plane is perpendicular to a base but only intersects the prism at an edge.

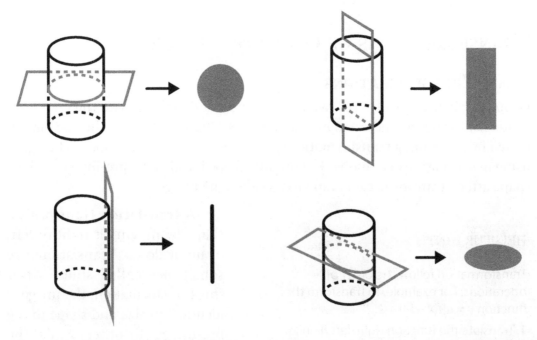

Figure 3.18. Plane Intersecting a Cylinder

SAMPLE QUESTIONS

17) **What would be the area of the shape that results when a cube with side length 5 cm is intersected by a plane that intersects the cube along one of its edges and proceeds along the cube's diagonal, cutting the cube into two equal pieces?**

Answer:

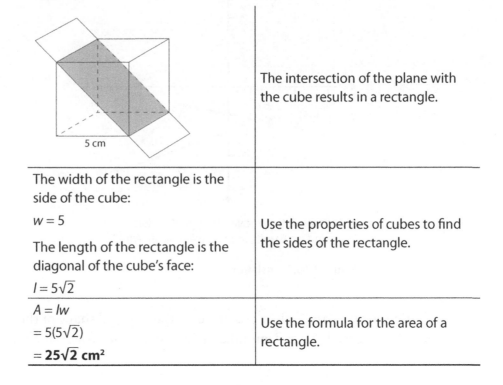

	The intersection of the plane with the cube results in a rectangle.
The width of the rectangle is the side of the cube: $w = 5$ The length of the rectangle is the diagonal of the cube's face: $l = 5\sqrt{2}$	Use the properties of cubes to find the sides of the rectangle.
$A = lw$ $= 5(5\sqrt{2})$ $= \mathbf{25\sqrt{2}}$ **cm²**	Use the formula for the area of a rectangle.

TRANSFORMATIONS OF GEOMETRIC FIGURES

BASIC TRANSFORMATIONS

Geometric figures are often drawn in the coordinate *xy*-plane, with the vertices or centers of the figures indicated by ordered pairs. These shapes can then be manipulated by performing **transformations**, which alter the size or shape of the figure using mathematical operations. The original shape is called the **pre-image**, and the shape after a transformation is applied is called the **image**.

HELPFUL HINT

Transformation follow the order of operations. For example, to transform the function $y = a[f(x - h)] + k$:

1. Translate the function right/left *h* units.
2. Dilate the function by the scale factor *a*.
3. Reflect the graph if $a < 0$.
4. Translate the function up/down *k* units.

A **translation** transforms a shape by moving it right or left, or up or down. Translations are sometimes called slides. After this transformation, the image is identical in size and shape to the pre-image. In other words, the image is **congruent**, or identical in size, to the pre-image. All corresponding pairs of angles are congruent, and all corresponding side lengths are congruent.

Translations are often in brackets: (x,y). The first number represents the change in the *x* direction (left/right), while the second number shows the change in the *y* direction (up/down).

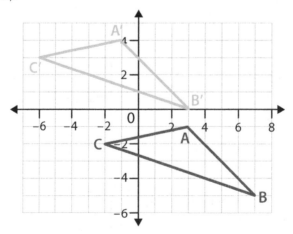

The translation moved triangle ABC left 4 units and up 6 units to produce triangle A'B'C'.

Figure 3.19. Translation

Similarly, rotations and reflections preserve the size and shape of the figure, so congruency is preserved. A **rotation** takes a pre-image and rotates it about a fixed point (often the origin) in the plane. Although the position or orientation of the shape changes, the angles and side lengths remain the same.

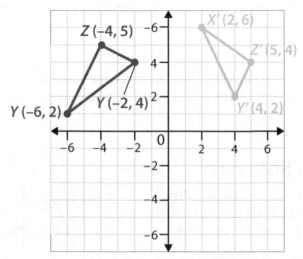

The triangle XYZ is rotated 90 in the clockwise direction about the origin (0, 0).

Figure 3.20. Rotation

A **reflection** takes each point in the pre-image and flips it over a point or line in the plane (often the *x*- or *y*-axis, but not necessarily). The image is congruent to the pre-image. When a figure is flipped across the *y*-axis, the signs of all *x*-coordinates will change. The *y*-coordinates change sign when a figure is reflected across the *x*-axis.

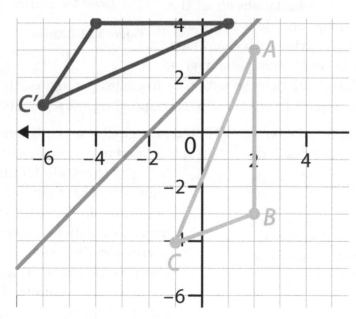

Figure 3.21. Reflection

18) If quadrilateral *ABCD* has vertices *A* (–6, 4), *B* (–6, 8), *C* (2, 8), and *D* (4, –4), what are the new vertices if *ABCD* is translated 2 units down and 3 units right?

Answer:

Translating two units down decreases each *y*-value by 2, and moving 3 units to the right increases the *x*-value by 3. The new vertices are *A* (–3, 2), *B* (–3, 6), *C* (5, 6), and *D* (7, –6).

DILATIONS AND SIMILARITY

A **dilation** increases (or decreases) the size of a figure by some **scale factor**. Each coordinate of the points that make up the figure is multiplied by the same factor. If the factor is greater than 1, multiplying all the factors enlarges the shape; if the factor is less than 1 (but greater than 0), the shape is reduced in size.

In addition to the scale factor, a dilation needs a **center of dilation**, which is a fixed point in the plane about which the points are multiplied. Usually, but not always, the center of dilation is the origin (0,0). For dilations about the origin, the image coordinates are calculated by multiplying each coordinate by the scale factor **k**. Thus, point (**x**, **y**) →

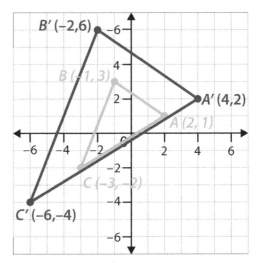

The triangle ABC is dilated by the scale factor 2 to produce triangle A'B'C'.

Figure 3.22. Dilation

(**kx,ky**). Although dilations do not result in congruent figures, the orientation of the figure is preserved; consequently, corresponding line segments will be parallel.

Importantly, dilations do NOT create images that are congruent to the original because the size of each dimension is increased or decreased (the only exception being if the scale factor is exactly 1). However, the shape of the figure is maintained. The corresponding angle measures will be congruent, but the corresponding side lengths will be *proportional*. In other words, the image and pre-image will be **similar** shapes (described with the symbol ~).

19) If quadrilateral *ABCD* has vertices *A* (–6, 4), *B* (–6, 8), *C* (2, 8), and *D* (4, –4), what are the new vertices if *ABCD* is increased by a factor of 5 about the origin?

Answer:

Multiply each point by the scale factor of 5 to find the new vertices: ***A* (–30, 20), *B* (–30, 40), *C* (10, 40), and *D* (20, –20).**

TRANSFORMING COORDINATES

Transformations in a plane can actually be thought of as functions. An input pair of coordinates, when acted upon by a transformation, results in a pair of output coordinates. Each point is moved to a unique new point (a one-to-one correspondence).

Table 3.4. How Coordinates Change for Transformations in a Plane

Type of Transformation	Coordinate Changes
Translation right *m* units and up *n* units	$(x, y) \rightarrow (x + m, y + n)$
Rotations about the origin in positive (counterclockwise) direction	
Rotation 90°	$(x, y) \rightarrow (-y, x)$
Rotation 180°	$(x, y) \rightarrow (-x, -y)$
Rotation 270°	$(x, y) \rightarrow (y, -x)$
Reflections about the	
***x*-axis**	$(x, y) \rightarrow (x, -y)$
***y*-axis**	$(x, y) \rightarrow (-x, y)$
line ***y* = *x***	$(x, y) \rightarrow (y, x)$
Dilations about the origin by a factor of *k*	
0 < *k* < 1 → size reduced	$(x, y) \rightarrow (kx, ky)$
***k* > 1 → size enlarged**	

20) If quadrilateral *ABCD* has vertices *A* (–6, 4), *B* (–6, 8), *C* (2, 8), and *D* (4, –4), what are the new vertices if *ABCD* is rotated 270° and then reflected across the *x*-axis?

Answer:

When a figure is rotated 270°, the coordinates change: $(a, b) \rightarrow (b, -a)$. After the rotation, the new coordinates are (4, 6), (8, 6), (8, –2), and (–4, –4). Reflecting across the *x*-axis requires that every *y*-value is multiplied by –1 to arrive at the completely transformed quadrilateral with vertices of (4, –6), (8, –6), (8, 2), and (–4, 4).

21) Triangle *ABC* with coordinates $(2, 8)$, $(10, 2)$, and $(6, 8)$ is transformed in the plane as shown in the diagram. What transformations result in the image triangle *A'B'C'*?

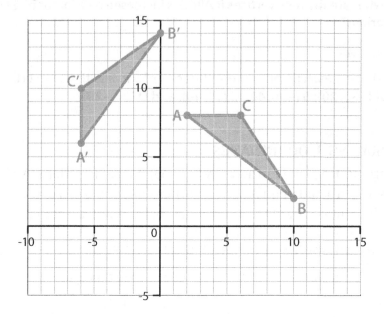

Answer:

Since the orientation of the triangle is different from the original, it must have been rotated. A counterclockwise rotation of 90° about the point *A* $(2, 8)$ results in a triangle with the same orientation. Then the triangle must be translated to move it to the image location. Pick one point, say *A*, and determine the translation necessary to move it to point *A'*. In this case, each point on the pre-image must be translated 8 units left and 2 units down, or $(-8, -2)$. (Note that this is one of many possible answers.)

TRIANGLE CONGRUENCE AND SIMILARITY

CONGRUENCE AND SIMILARITY IN TRIANGLES

Congruence and similarity in triangles is governed by a set of theorems that make it easy to determine the relationship between two triangles. These theorems can be used by looking at the number and location of congruent sides and angles shared by the triangles.

Table 3.5. Triangle Congruence

Triangles are congruent if...	
SSS	all three corresponding sides are congruent.
SAS	two sides and the included angle are congruent.
AAS	two angles and one side are congruent.
ASA	two angles and the included side are congruent.

Triangles are similar if...	
AA	two angles (and thus the third) are congruent.
SAS	two sides are proportional and included angle is congruent.
SSS	the ratio between all three corresponding sides is constant.

A note of caution about the SSA case (also known as the ASS case): Two triangles that have congruent (or proportional) sides and one congruent (but not included) angle are NOT necessarily congruent (or similar). This is because, given two side lengths and a nonincluded angle, it is often possible to draw an acute triangle and an obtuse triangle. Thus, there is NOT an SSA congruence theorem.

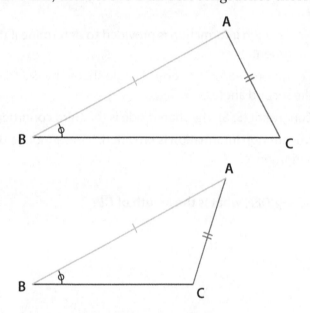

Figure 3.23. Two Possible Triangles Formed from
Two Sides and One Non-Included Angle

Right triangles, in general, have a couple of simpler similarity and congruence theorems. Since any pair of right triangles already has one angle that is the same (the right angle), any two pairs of corresponding sides that are either congruent or proportional will guarantee congruence or similarity between the triangles, respectively.

Table 3.6. Right Triangle Congruence

Right triangles are congruent if...	
HL	the hypotenuse and one leg are congruent.
LL	the two pairs of corresponding legs are congruent.

Right triangles are similar if...	
HL	the hypotenuse and one leg are proportional.
LL	two pairs of corresponding legs are similar.

SAMPLE QUESTIONS

22) Determine whether the following sets of triangles are congruent. If they are, state why.

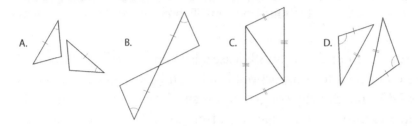

Answer:

A.   Not enough information is provided to determine if the triangles are congruent.

B.   Congruent (ASA). The vertical angle shared by the triangles provides the second angle.

C.   Congruent (SSS). The shared side is the third congruent side.

D.   Not enough information is provided: SSA cannot be used to prove congruence.

23) If $\triangle ABC \sim \triangle DEF$, what is the length of $\overline{DE}$?

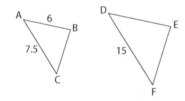

Answer:

Set up a proportion among the sides:

$$\frac{AB}{AC} = \frac{ED}{DF}$$

$$\frac{6}{7.5} = \frac{DE}{15}$$

$DE = 12$

COORDINATE GEOMETRY

Coordinate geometry combined with transformations can be useful for proving many geometric theorems. Many other theorems about polygons can be proven using coordinate geometry by computing distances, slopes, and midpoints of figures in the plane.

THE DISTANCE AND MIDPOINT FORMULAS

The distance formula finds the distance of a line drawn between two points that terminates at those two points:

$$d = \sqrt{(x_2 - x_1)^2 + (y_2 - y_1)^2}$$

The distance formula resembles the Pythagorean theorem because it is essentially finding the hypotenuse of the right triangle with legs of length $\Delta x = x_2 - x_1$ and $\Delta y = y_2 - y_1$.

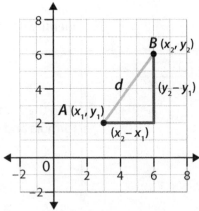

Figure 3.24. The Distance Formula

The midpoint formula finds the coordinates of a point exactly in the middle of two other points. To find the midpoint, average the x values and then average the y values. These are the coordinates of the midpoint:

$$\text{Midpoint } M = \left(\frac{x_1 + x_2}{2}, \frac{y_1 + y_2}{2} \right)$$

SAMPLE QUESTIONS

24) If $(-3, 8)$ is the midpoint of segment $\overline{AB}$ and point A is at $(-10, -17)$, what are the coordinates of point B?

Answer:

$A = (x_1, y_1) = (-10, -17)$	
$B = (x_2, y_2)$	Identify the given variables.
$M = (-3, 8)$	
$M_x = \frac{x_2 - x_1}{2}$	
$-3 = \frac{x_2 - (-10)}{2}$	Use the midpoint formula to find point B.
$x_2 = 4$	
$M_y = \frac{y_2 - y_1}{2}$	

$$8 = \frac{y_2 - (-17)}{2}$$

$$y_2 = 33$$

B = (4, 33)

25) **Find the distance between the points (−10, 50) and (50, 10).**

Answer:

$(x_1, y_1) = (-10, 50)$ $(x_2, y_2) = (50, 10)$	Identify the given variables.
$d = \sqrt{(x_2 - x_1)^2 + (y_2 - y_1)^2}$ $d = \sqrt{50 - (-10)^2 + (10 - 50)^2}$ $d = \sqrt{(60)^2 + (-40)^2}$ $d = \sqrt{3600 + 1600}$ $d = \sqrt{5200}$ **$d \approx 72.11$**	Plug these values into the distance formula and solve.

COORDINATE PROOFS

One method of proving properties about a figure is to use the coordinate plane to define vertices or other important points, and then use slope, the distance formula, the midpoint formula, or other knowledge about the coordinates.

Given the coordinates on the figure (the diagram is not to scale) it is possible to prove that the quadrilateral is a parallelogram. There are many ways to do this. For instance, to prove that opposite sides are both parallel, find the slopes of all four sides:

$$m_{OA} = \frac{b - 0}{a - 0} = \frac{b}{a} \qquad m_{AB} = \frac{b - b}{2a - a} = 0$$

$$m_{BC} = \frac{b - 0}{2a - a} = \frac{b}{a} \qquad m_{OC} = \frac{0 - 0}{a - 0} = 0$$

Since the slopes of $\overline{OA}$ and $\overline{BC}$ are the same, the segments are parallel, and since $\overline{AB}$ and $\overline{OC}$ both have slope 0, they are parallel; thus the shape is a parallelogram.

It is also possible to prove that the diagonals bisect each other by finding the midpoints of the diagonals. Diagonal $\overline{AC}$ has midpoint $\left(\frac{a + a}{2}, \frac{b + 0}{2}\right)$, while diagonal $\overline{OB}$ has midpoint $\left(\frac{2a + 0}{2}, \frac{b + 0}{2}\right)$. Since these both simplify to the same point $(a, \frac{b}{2})$, the diagonals must intersect there. In addition, because these are midpoints, the diagonals are bisecting each other.

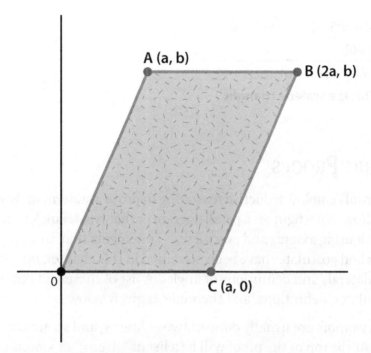

Figure 3.25. Quadrilateral on the Coordinate Plane

SAMPLE QUESTION

26)   **Classify the triangle below as scalene, isosceles, or equilateral.**

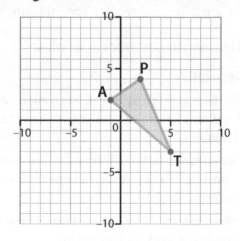

Answer:

$A = (-1, 2)$

$P = (2, 4)$

$T = (5, -3)$

$AP = \sqrt{2 - (-1)^2 + (4 - 1)^2}$

$= 3\sqrt{2}$

$PT = \sqrt{(5 - 2)^2 + (-3 - 4)^2}$

$= \sqrt{58}$

Use the distance formula to find the lengths of each side of the triangle.

$$AT = \sqrt{(5 - (-1)^2) + (-3 - 2)^2}$$
$$= \sqrt{61}$$

$3\sqrt{2} \neq \sqrt{58} \neq \sqrt{61}$ **This is a scalene triangle.**	Classify the triangle.

GEOMETRIC PROOFS

Geometric proofs employ deductive reasoning to prove a statement. For example, a proof might show that a figure is a parallelogram or that two triangles are congruent. Proofs are built using axioms and postulates—statements that are accepted as true. Many axioms and postulates have been presented in this chapter, including properties of quadrilaterals and definitions of angles. A list of these and other important axioms, postulates, definitions, and theorems is given below.

Geometry proofs are usually done in two columns, and so are called **two-column proofs**. At the top of the proof will be a list of "givens," or statements that are true because they are given in the problem. The last row of the proof will be the statement that is being proved true. In between will be a list of statements that support the conclusion. The left column includes a specific statement, such as m∠A $\cong m\angle B$, and the right column includes the axiom or postulate that is being used to declare that statement true. These middle steps should use axioms and postulates to build on the given information and reach a specific conclusion.

Table 3.7. Common Geometric Postulates: Angles

Angles	Description
Parallel lines	Two lines crossed by a third line (called a transversal) are parallel if and only if the alternate interior (or exterior) angles are congruent.
	Two lines crossed by a transversal are parallel if and only if same-side (of the transversal) interior (or exterior) angles are supplementary.
	Two lines crossed by a transversal are parallel if and only if corresponding angles are congruent.
Linear pairs	Two angles of a linear pair are supplementary (180°).
Vertical pairs	Two angles of a vertical pair are congruent.

Table 3.8. Common Geometric Postulates: Triangles

Triangles	Two triangles are congruent (or similar) if ...
Angle-side-angle congruence (ASA)	they have two pairs of corresponding angles that are congruent and the sides between the corresponding angles are congruent.
Side-angle-side congruence (SAS)	they have two pairs of corresponding sides that are congruent and the corresponding angles between the sides are congruent.

Triangles	Two triangles are congruent (or similar) if ...
Side-side-side congruence (SSS)	all three corresponding pairs of sides are congruent.
Angle-angle-side congruence (AAS)	they have two pairs of congruent angles and a pair of congruent sides not in-between those angles.
Angle-angle similarity (AA)	they have two pairs of corresponding congruent angles.
Side-angle-side similarity (SAS)	they have two pairs of corresponding sides that are proportional and the corresponding angles between the sides are congruent.
Side-side-side similarity (SSS)	all three corresponding pairs of the triangles are proportional.

Triangles	Description
Corresponding parts of congruent triangles are congruent (CPCTC)	after proving two triangles are congruent using above theorems, it is known that all corresponding parts of the triangles are congruent.
Midsegment theorem	a line segment connecting the midpoints of two legs of a triangle is parallel to the third leg and half as long (x).

Table 3.9. Common Geometric Postulates: Quadrilaterals

Quadrilaterals	Description
Parallelograms	If a quadrilateral is a parallelogram, its opposite sides are congruent and its diagonals create two congruent triangles.
Rectangle	A parallelogram with one right angle must be a rectangle.
Trapezoid	A trapezoid may have only one pair of congruent lines.

SAMPLE QUESTIONS

27) Prove: $\overline{AB} \cong \overline{CD}$ and $\overline{AD} \cong \overline{BC}$ in the parallelogram below.

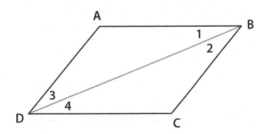

Answer:

Statement	Axiom/Postulate
$\overline{AB} \parallel \overline{CD}$ and $\overline{AD} \parallel \overline{BC}$	Opposite sides of a parallelogram are parallel.
$m\angle 1 \cong m\angle 4$ and $m\angle 2 \cong m\angle 3$	Alternate interior angles formed by parallel lines and a transversal are congruent.
$\overline{BD} \cong \overline{DB}$	Reflexive property
$\triangle DAB \cong \triangle BCD$	AAS congruence
$\therefore$ **$AB \cong CD$ and $AD \cong BC$**	Corresponding parts of congruent triangles are congruent (CPCTC).

CONSTRUCTIONS OF GEOMETRIC FIGURES

Constructing geometric shapes, including bisectors and tangent lines, helps clarify the relationships between the figures. Constructions are made using only a **compass** with a point and pencil, and a **straight edge**.

CONSTRUCTING AN EQUILATERAL TRIANGLE

To construct an equilateral triangle, begin by setting the compass to the length of each side of the triangle. Next, define a point A. Place the point of the compass at A, and then draw two arcs in the approximate locations where the other vertices should be. Then, define a point B anywhere along one of these arcs. Place the point of the compass on B and make an intersecting arc with the remaining arc. Point C is the intersection of those arcs. Use a straight edge to connect the vertices to form an equilateral triangle.

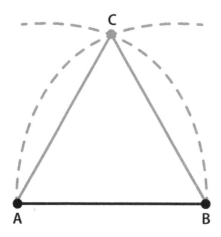

Figure 3.26. Constructing an Equilateral Triangle

CONSTRUCTING A PERPENDICULAR BISECTOR OF A SEGMENT

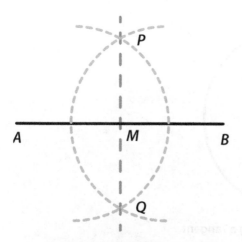

Figure 3.27. Constructing the Perpendicular Bisector of a Segment

Begin by drawing a line segment. Label it $\overline{AB}$. Put the point of the compass on point A, and extend the compass until it is more than half the length of the segment. Make an arc somewhere above the line segment. Then, make an arc somewhere below the line segment. Without changing the setting on the compass, place the point on point B, make an intersecting arc with the arc above the segment, and another intersecting the arc below the segment. Using a straight edge, the two intersecting points can be connected with a line, which is the perpendicular bisector of the segment. The point M, where the perpendicular bisector intersects the segment, is the midpoint of the segment.

CONSTRUCTING AN ANGLE BISECTOR

Begin drawing an angle with vertex A. Put the point of the compass on point A, and open the compass a bit. Make an arc along each of the rays of the angle. Then place the point of the compass at the intersection of one of the arcs and make another arc in the interior of the angle. Repeat, using the same compass setting, at the other intersection point, making sure to intersect the interior arcs. Connect the vertex of the angle to the intersection point of the interior arcs to create the angle bisector.

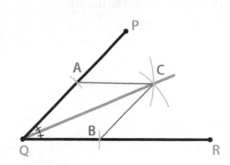

Figure 3.28. Constructing an Angle Bisector

CONSTRUCTING A TANGENT TO A CIRCLE

To construct a tangent to a circle, begin by using a compass to draw a circle, and then mark a point on the circle. Then draw an extended radius through the point into the exterior of the circle.

Next, construct a perpendicular line to this line. To do this, put the point of the compass on the point on the circle, and make an arc (a mark) anywhere on the radius. Without moving the point of the compass, swing the compass around and make another mark, the same distance away, at the intersection of the radial line on the exterior portion of the circle.

Next, place the point of the compass on one of these intersection points, widen the compass a bit, and make sweeping marks on either side of the radial line.

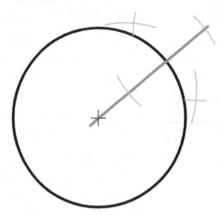

Figure 3.29. Constructing a Tangent to a Circle

Switch the point of the compass to the other radial line intersection mark, and make another set of sweeping marks to form X's on the exterior of the circle.

With a straight edge, connect the center points of the X's; the tangent line to the circle is constructed.

4

Statistics

tatistics is the study of **data**, which are simply sets of qualitative and quantitative values. These values are often the result of observations or measurements collected as part of experiments or surveys. The sections below discuss how to organize, analyze, and present data in a variety of ways.

DESCRIBING SETS OF DATA

MEASURES OF CENTRAL TENDENCY

Measures of central tendency help identify the center, or most typical, value within a data set. There are three such central tendencies that describe the "center" of the data in different ways. The **mean** is the arithmetic average and is found by dividing the sum of all measurements by the number of measurements. The mean of a population is written as μ and the mean of a sample is written as $\bar{x}$.

$$\text{population mean} = \mu = \frac{x_1 + x_2 + ...x_N}{N} = \frac{\Sigma x}{N} \qquad \text{sample mean} = \bar{x} = \frac{x_1 + x_2 + ...x_n}{n} = \frac{\Sigma x}{n}$$

The data points are represented by x's with subscripts; the sum is denoted using the Greek letter sigma (Σ); N is the number of data points in the entire population; and n is the number of data points in a sample set.

    The **median** divides the measurements into two equal halves. The median is the measurement right in the middle of an odd set of measurements or the average of the two middle

HELPFUL HINT

When the same value is added to each term in a set, the mean increases by that value and the standard deviation is unchanged.

When each term in a set is multiplied by the same value, both the mean and standard deviation will also be multiplied by that value.

numbers in an even data set. When calculating the median, it is important to order the data values from least to greatest before attempting to locate the middle value.

The **mode** is simply the measurement that occurs most often. There can be many modes in a data set, or no mode. Since measures of central tendency describe a *center* of the data, all three of these measures will be between the lowest and highest data values (inclusive).

Unusually large or small values, called **outliers**, will affect the mean of a sample more than the mode. If there is a high outlier, the mean will be greater than the median; if there is a low outlier, the mean will be lower than the median. When outliers are present, the median is a better measure of the data's center than the mean because the median will be closer to the terms in the data set.

SAMPLE QUESTIONS

1) **What is the mean of the following data set? {1000, 0.1, 10, 1}**

   Answer:

   Use the equation to find the mean of a sample:

   $$\frac{1000 + 0.1 + 10 + 1}{4} = \textbf{252.78}$$

2) **What is the median of the following data set? {1000, 10, 1, 0.1}**

   Answer:

   Since there are an even number of data points in the set, the median will be the mean of the two middle numbers. Order the numbers from least to greatest: 0.1, 1, 10, and 1000. The two middle numbers are 1 and 10, and their mean is:

   $$\frac{1 + 10}{2} = \textbf{5.5}$$

3) **Josey has an average of 81 on four equally weighted tests she has taken in her statistics class. She wants to determine what grade she must receive on her fifth test so that her mean is 83, which will give her a B in the course, but she does not remember her other scores. What grade must she receive on her fifth test?**

   Answer:

   Even though Josey does not know her test scores, she knows her average. Therefore it can be assumed that each test score was 81, since four scores of 81 would average to 81. To find the score, x, that she needs use the equation for the mean of a sample:

   $$\frac{4(81) + x}{5} = 83$$

   $$324 + x = 415$$

   $$x = \textbf{91}$$

MEASURES OF VARIATION

The values in a data set can be very close together (close to the mean), or very spread out. This is called the **spread** or **dispersion** of the data. There are a few **measures of variation** (or **measures of dispersion**) that quantify the spread within a data set. **Range** is the difference between the largest and smallest data points in a set:

$$R = largest\ data\ point - smallest\ data\ point$$

Notice range depends on only two data points (the two extremes). Sometimes these data points are outliers; regardless, for a large data set, relying on only two data points is not an exact tool.

The understanding of the data set can be improved by calculating **quartiles**. To calculate quartiles, first arrange the data in ascending order and find the set's median (also called quartile 2 or Q2). Then find the median of the lower half of the data, called quartile 1 (Q1), and the median of the upper half of the data, called quartile 3 (Q3). These three points divide the data into four equal groups of data (thus the word *quartile*). Each quartile contains 25% of the data.

Interquartile range (IQR) provides a more reliable range that is not as affected by extremes. IQR is the difference between the third quartile data point and the first quartile data point and gives the spread of the middle 50% of the data:

$$IQR = Q_3 - Q_1$$

A measure of variation that depends on the mean is **standard deviation**, which uses every data point in a set and calculates the average distance of each data point from the mean of the data. Standard deviation can be computed for an entire population (written σ) or for a sample of a population (written s):

$$\sigma = \sqrt{\frac{\Sigma(x_i - \mu)^2}{N}} \qquad s = \sqrt{\frac{\Sigma(x_i - \overline{x})^2}{n - 1}}$$

Thus, to calculate standard deviation, the difference between the mean and each data point is calculated. Each of these differences is squared (so that each is positive). The average of the squared values is computed by summing the squares and dividing by N or $(n - 1)$. Then the square root is taken, to "undo" the previous squaring.

The **variance** of a data set is simply the square of the standard variation:

> **HELPFUL HINT**
>
> Standard deviation and variance are also affected by extreme values. Though much simpler to calculate, interquartile range is the more accurate depiction of how the data is scattered when there are outlier values.

$$V = \sigma^2 = \frac{1}{N} \sum_{i=1}^{N} (x_i - \mu)^2$$

Variance measures how narrowly or widely the data points are distributed. A variance of zero means every data point is the same; a large variance means the data is widely spread out.

SAMPLE QUESTIONS

4)   **What are the range and interquartile range of the following set? {3, 9, 49, 64, 81, 100, 121, 144, 169}**

Answer:

R = largest point − smallest point $= 169 - 3$ $= \mathbf{166}$	Use the equation for range.
3 9 → Q1 = $\frac{49 + 9}{2}$ = 29 49 64 81 → Q2 100 121 → Q3 = $\frac{121 + 144}{2}$ =132.5 144 169	Place the terms in numerical order and identify Q1, Q2, and Q3.
IQR = Q3 − Q1 $= 132.5 - 29$ $= \mathbf{103.5}$	Find the IQR by subtracting Q1 from Q3.

5)   **In a group of 7 people, 1 person has no children, 2 people have 1 child, 2 people have 2 children, 1 person has 5 children, and 1 person has 17 children. To the nearest hundredth of a child, what is the standard deviation in this group?**

Answer:

{0, 1, 1, 2, 2, 5, 17}	Create a data set out of this scenario.
$\mu = \dfrac{x_1 + x_2 + ...x_N}{N} = \dfrac{\Sigma x}{N}$ $\mu = \dfrac{0 + 1 + 1 + 2 + 2 + 5 + 17}{7} = 4$	Calculate the population mean.
$(0 - 4)^2 = (-4)^2 = 16$ $(1 - 4)^2 = (-3)^2 = 9$ $(1 - 4)^2 = (-3)^2 = 9$ $(2 - 4)^2 = (-2)^2 = 4$ $(2 - 4)^2 = (-2)^2 = 4$ $(5 - 4)^2 = (1)^2 = 1$ $(17 - 4)^2 = (13)^2 = 169$	Find the square of the difference of each term and the mean $(x_i - \mu)^2$.

$$\sigma = \sqrt{\frac{\Sigma(x_i - \mu)^2}{N}}$$

$$\sigma = \sqrt{\frac{212}{7}} = \sqrt{30.28} = \mathbf{5.50}$$

Plug the sum (Σ) of these squares, 212, into the standard deviation formula.

Box Plots

A box plot depicts the median and quartiles along a scaled number line. It is meant to summarize the data in a visual manner and emphasize central trends while decreasing the pull of outlier data. To construct a box plot:

1. Create a number line that begins at the lowest data point and terminates at the highest data point.

2. Find the quartiles of the data. Create a horizontal rectangle (the "box") whose left border is Q_1 and right border is Q_3.

3. Draw a vertical line within the box to mark the median.

4. Draw a horizontal line going from the left edge of the box to the smallest data value.

5. Draw a horizontal line going from the right edge of the box to the largest data value.

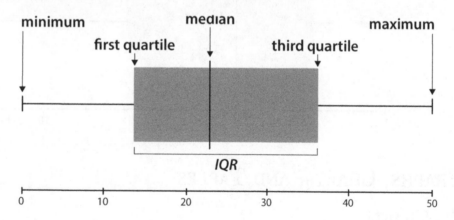

Figure 4.1. Box Plot

When reading a box plot, the following stands out:

▸ Reading from left to right: the horizontal line (whisker) shows the spread of the first quarter; the box's left compartment shows the spread of the second quarter; the box's right compartment shows the spread of

HELPFUL HINT

Box plots are also known as box-and-whisker plots, because if they are drawn correctly the two horizontal lines look like whiskers.

the third quarter; and the right horizontal line shows the spread of the fourth quarter.

▶ The length of the box is the IQR, or the middle 50% of the data.

▶ Each of the four pieces (the whiskers and two pieces in the box) represent 25% of the data.

▶ The horizontal lines show by their length whether the data higher or lower than the middle 50% is prominent.

SAMPLE QUESTION

6) **A recent survey asked 8 people how many pairs of shoes they wear per week. Their answers are in the following data set: {1, 3, 5, 5, 7, 8, 12}. Construct a box plot from this data.**

Answer:

Create a number line that begins at 1 and ends at 12. Q_1 is 3, the median (Q_2) is 5, and Q_3 is 8. A rectangle must be drawn whose length is 5 and that borders on Q_1 and Q_3. Mark the median of 5 within the rectangle. Draw a horizontal line going left to 1. Draw a horizontal line going right to 12.

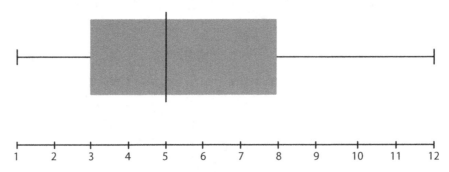

GRAPHS, CHARTS, AND TABLES

PIE CHARTS

A pie chart simply states the proportion of each category within the whole. To construct a pie chart, the categories of a data set must be determined. The frequency of each category must be found and that frequency converted to a percent of the total. To draw the pie chart, determine the angle of each slice by multiplying the percentage by 360°.

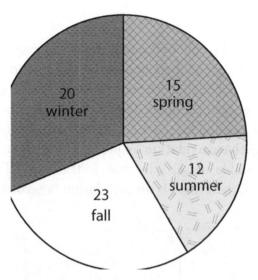

Figure 4.2. Pie Chart

SAMPLE QUESTION

7)   A firm is screening applicants for a job by education-level attainment.
     There are 125 individuals in the pool: 5 have a doctorate, 20 have a
     master's degree, 40 have a bachelor's degree, 30 have an associate degree,
     and 30 have a high school degree. Construct a pie chart showing the
     highest level of education attained by the applicants.

Answer:

Create a frequency table to find the percentages and angle measurement for
each category.

Category	Frequency	Percent	Angle Measure
High School	30	24%	86.4
Associate	30	24%	86.4
Bachelor's	40	32%	115.2
Master's	20	16%	57.6
Doctorate	5	4%	14.4

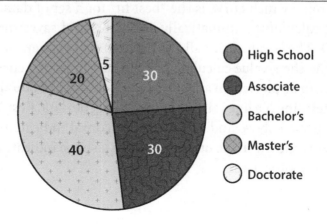

SCATTER PLOTS

A scatter plot is displayed in the first quadrant of the *xy*-plane where all numbers are positive. Data points are plotted as ordered pairs, with one variable along the horizontal axis and the other along the vertical axis. Scatter plots can show if there is a correlation between two variables. There is a **positive correlation** (expressed as a positive slope) if increasing one variable appears to result in an increase in the other variable. A **negative correlation** (expressed as a negative slope) occurs when an increase in one variable causes a decrease in the other. If the scatter plot shows no discernible pattern, then there is no correlation (a zero, mixed, or indiscernible slope).

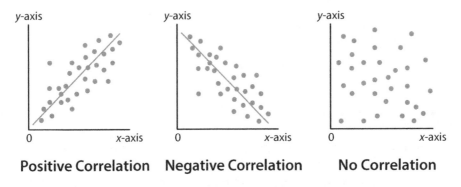

Positive Correlation   Negative Correlation   No Correlation

Figure 4.3. Scatter Plots and Correlation

Calculators or other software can be used to find the linear regression equation, which describes the general shape of the data. Graphing this equation produces the regression line, or line of best fit. The equation's **correlation coefficient** (*r*) can be used to determine how closely the equation fits the data. The value of *r* is between −1 and 1. The closer *r* is to 1 (if the line has a positive slope) or −1 (if the line has a negative slope), the better the regression line fits the data. The closer the *r* value is to 0, the weaker the correlation between the line and the data. Generally, if the absolute value of the correlation coefficient is 0.8 or higher, then it is considered to be a strong correlation, while an |*r*| value of less than 0.5 is considered a weak correlation.

To determine which curve is the "best fit" for a set of data, **residuals** are calculated. The calculator automatically calculates and saves these values to a list called RESID. These values are all the differences between the actual *y*-value of data points and the *y*-value calculated by the best-fit line or curve for that *x*-value. These values can be plotted on an *xy*-plane to produce a **residual plot**. The residual plot helps determine if a line is the best model for the data. Residual points that are randomly dispersed above and below the horizontal indicate that a linear model is appropriate, while a *u* shape or upside-down *u* shape indicate a nonlinear model would be more appropriate.

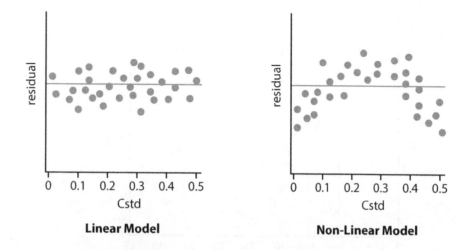

Figure 4.4. Residual Plots

Once a best-fit line is established, it can be used to estimate output values given an input value within the domain of the data. For a short extension outside that domain, reasonable predictions may be possible. However, the further from the domain of the data the line is extended, the greater the reduction in the accuracy of the prediction.

It is important to note here that just because two variables have a strong positive or negative correlation, it cannot necessarily be inferred that those two quantities have a *causal* relationship—that is, that one variable changing *causes* the other quantity to change. There are often other factors that play into their relationship. For example, a positive correlation can be found between the number of ice cream sales and the number of shark attacks at a beach. It would be incorrect to say that selling more ice cream *causes* an increase in shark attacks. It is much more likely that on hot days more ice cream is sold, and many more people are swimming, so one of them is more likely to get attacked by a shark. Confusing correlation and causation is one of the most common statistical errors people make.

> **HELPFUL HINT**
>
> A graphing calculator can provide the regression line, *r* value, and residuals list.

SAMPLE QUESTION

8) Based on the scatter plot on the following page, where the *x*-axis represents hours spent studying per week and the *y*-axis represents the average percent grade on exams during the school year, is there a correlation between the amount of studying for a test and test results?

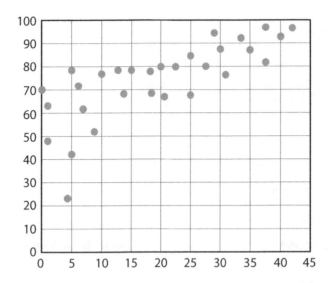

Answer:

There is a somewhat weak positive correlation. As the number of hours spent studying increases, the average percent grade also generally increases.

LINE GRAPHS

Line graphs are used to display a relationship between two variables, such as change over time. Like scatter plots, line graphs exist in quadrant 1 of the *xy*-plane. Line graphs are constructed by graphing each point and connecting each point to the next consecutive point by a line. To create a line graph, it may be necessary to consolidate data into single bivariate data points. Thus, a line graph is a function, with each *x*-value having exactly one *y*-value, whereas a scatter plot may have multiple *y*-values for one *x*-value.

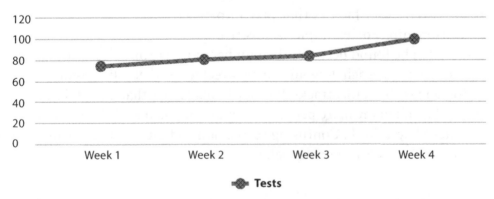

Figure 4.5. Line Graph

SAMPLE QUESTION

9)   Create a line graph based on the following survey values, where the first column represents an individual's age and the other represents that individual's reported happiness level on a 20-point scale (0 being the least

happy that person has been and 20 being the happiest). Then interpret the resulting graph to determine whether the following statement is true or false: *On average, middle-aged people are less happy than young or older people are.*

Age	Happiness
12	16
13	15
20	18
15	12
40	5
17	17
18	18
19	15
42	7
70	17
45	10
60	12
63	15
22	14
27	15
36	12
33	10
44	8
55	10
80	10
15	13
40	8
17	15
18	17
19	20
22	16
27	15
36	9
33	10
44	6

Answer:

To construct a line graph, the data must be ordered into consolidated categories by averaging the data of people who have the same age so that the data is one-to-one. For example, there are 2 twenty-two-year-olds who are reporting. Their average happiness level is 15. When all the data has been consolidated and ordered from least to greatest, the table and graph below can be presented.

Age	Happiness
12	16
13	15
15	12.5
17	16
18	17.5
19	17.5
20	18
22	15
27	15
33	10
36	10.5
40	6.5
42	7
44	7
45	10
55	10
60	12
63	15
70	17
80	10

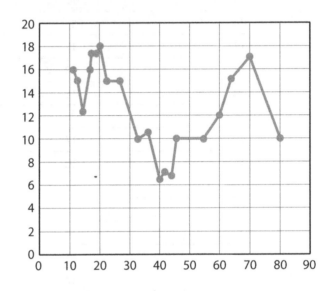

Average Happiness Rating Versus Age

The statement that, on average, middle-aged people are less happy than young or older people appears to be true. According to the graph, people in their thirties, forties, and fifties are less happy than people in their teens, twenties, sixties, and seventies.

BAR GRAPHS

Bar graphs compare differences between categories or changes over a time. The data is grouped into categories or ranges and represented by rectangles. A bar graph's rectangles can be vertical or horizontal, depending on whether the dependent variable is placed on the *x*- or *y*-axis. Instead of the *xy*-plane, however, one axis is made up of categories (or ranges) instead of a numeric scale. Bar graphs are useful because the differences between categories are easy to see: the height or length of each bar shows the value for each category.

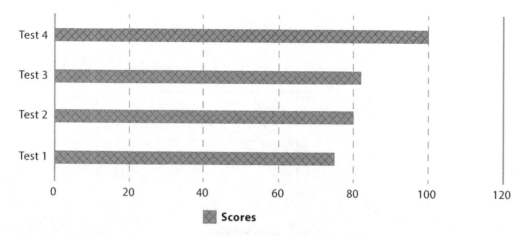

Figure 4.6. Bar Graph

SAMPLE QUESTION

10) A company X had a profit of $10,000 in 2010, $12,000 in 2011, $15,600 in 2012, and $20,280 in 2013. Create a bar graph displaying the profit from each of these four years.

Answer:

Place years on the independent axis, and profit on the dependent axis, and then draw a box showing the profit for each year.

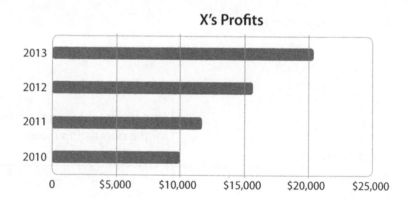

STEM-AND-LEAF PLOTS

Stem-and-leaf plots are ways of organizing large amounts of data by grouping it into classes. All data points are broken into two parts: a stem and a leaf. For instance, the number 512 might be broken into a stem of 5 and a leaf of 12. All data in the 500 range would appear in the same row (this group of data is a class). Usually a simple key is provided to explain how the data is being represented. For instance, 5|12 = 512 would show that the stems are representing hundreds. The advantage of this display is that it shows general density and shape of the data in a compact display, yet all original data points are preserved and available. It is also easy to find medians and quartiles from this display.

Stem	Leaf
0	5
1	6, 7
2	8, 3, 6
3	4, 5, 9, 5, 5, 8, 5
4	7, 7, 7, 8
5	5, 4
6	0

Figure 4.7. Stem and Leaf Plot

SAMPLE QUESTION

11) The table gives the weights of wrestlers (in pounds) for a certain competition. What is the mean, median, and IQR of the data?

2	05, 22, 53, 40
3	07, 22, 29, 45, 89, 96, 98
4	10, 25, 34
6	21

Key: 2|05 = 205 pounds

Answer:

$\mu = \frac{\sum x}{N}$ $= \frac{5281}{15}$ $= \textbf{353.1 lbs.}$	Find the mean using the equation for the population mean.
Q1 = 253 Q2 = 345 Q3 = 410 IQR = 410 − 253 = 157 **The median is 345 lbs.** **The IQR is 157 lbs.**	Find the median and IQR by counting the leaves and identifying Q1, Q2, and Q3.

FREQUENCY TABLES AND HISTOGRAMS

The frequency of a particular data point is the number of times that data point occurs. Constructing a frequency table requires that the data or data classes be arranged in ascending order in one column and the frequency in another column.

A histogram is a graphical representation of a frequency table used to compare frequencies. A histogram is constructed in quadrant 1 of the *xy*-plane, with data in each equal-width class presented as a bar and the height of each bar representing the frequency of that class. Unlike bar graphs, histograms cannot have gaps between bars. A histogram is used to determine the distribution of data among the classes.

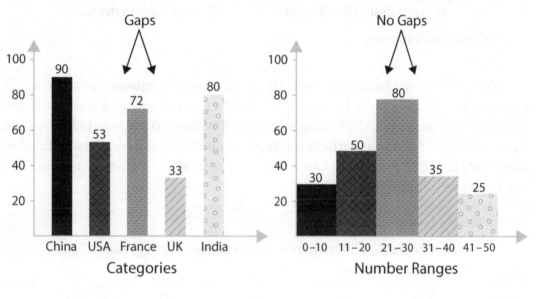

Bar Chart   **Histogram**

Figure 4.8. Bar Charts vs. Histograms

Histograms can be symmetrical, skewed left or right, or multimodal (data spread around). Note that **skewed left** means the peak of the data is on the *right*, with a tail to the left, while **skewed right** means the peak is on the *left*, with a tail to the right. This seems counterintuitive to many; the "left" or "right" always refers to the tail of the data. This is because a long tail to the right, for example, means there are high outlier values that are skewing the data to the right.

A **two-way frequency table** compares **categorical data** (data in more than one category) of two related variables (bivariate data). Two-way frequency tables are also called **contingency tables** and are often used to analyze survey results. One category is displayed along the top of the table and the other category down along the side. Rows and columns are added and the sums appear at the end of the row or column. The sum of all the row data must equal the sum of all the column data.

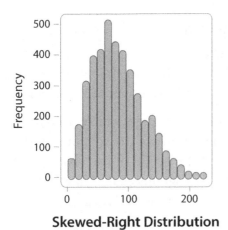 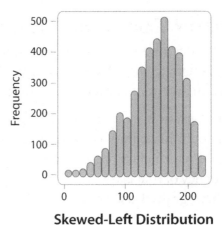

Skewed-Right Distribution   Skewed-Left Distribution

Figure 4.9. Histograms

From a two-way frequency table, the **joint relative frequency** of a particular category can be calculated by taking the number in the row and column of the categories in question and dividing by the total number surveyed. This gives the percent of the total in that particular category. Sometimes the **conditional relative frequency** is of interest. In this case, calculate the relative frequency confined to a single row or column.

Students by Grade and Gender

	9th grade	10th grade	11th grade	12th grade	Total
Male	57	63	75	61	256
Female	54	42	71	60	227
Total	111	105	146	121	483

Figure 4.10. Two-Way Frequency Table

SAMPLE QUESTIONS

12) A café owner tracked the number of customers he had over a twelve-hour period in the frequency table below. Display the data in a histogram and determine what kind of distribution there is in the data.

Time	Number of Customers
6 a.m. – 8 a.m.	5
8 a.m. – 9 a.m.	6
9 a.m. – 10 a.m.	5
10 a.m. – 12 p.m.	23
12 p.m. – 2 p.m.	24
2 p.m. – 4 p.m.	9
4 p.m. – 6 p.m.	4

Answer:

Since time is the independent variable, it is on the *x*-axis and the number of customers is on the *y*-axis. For the histogram to correctly display data continuously, categories on the *x*-axis must be equal 2-hour segments. The 8 a.m. – 9 a.m. and 9 a.m. – 10 a.m. categories must be combined for a total of 11 customers in that time period. Although not perfectly symmetrical, the amount of customers peaks in the middle and is therefore considered symmetrical.

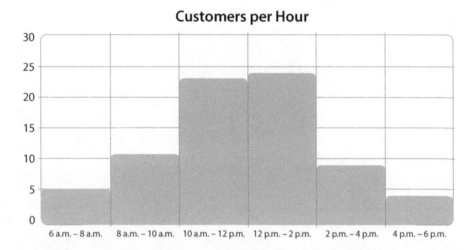

13) Cineflix movie theater polled its moviegoers on a weeknight to determine their favorite type of movie. The results are in the two-way frequency table below.

Moviegoers	Comedy	Action	Horror	Totals
Male	15	24	21	60
Female	8	18	17	43
Totals	23	42	38	103

Determine whether each of the following statements is true or false.

A. Action films are the most popular type of movie

B. About 1 in 5 moviegoers prefers comedy films

C. Men choose the horror genre more frequently than women do

Answer:

A. **True.** More people (42) chose action movies than comedy (23) or horror (38).

B. **True.** Find the ratio of total number of people who prefer comedy to total number of people. $\frac{23}{103} = 0.22$; 1 in 5 is 20% so 22% is about the same.

C. **False.** The percentage of men who choose horror is less than the percentage of women who do.

 part = number of men who prefer horror =21

whole = *number of men surveyed* = 60

$$percent = \frac{part}{whole}$$

$$= \frac{21}{60} = 0.35 = 35\%$$

part = *number of women who prefer horror* = 17

whole = *number of women surveyed* = 43

$$percent = \frac{part}{whole}$$

$$= \frac{17}{43} = 0.40 = 40\%$$

5

Logic and Probability

This chapter starts by giving the basic rules and processes for applying mathematical logic, such as creating truth tables and using set theory. Understanding how logic is used in math is important when studying probability, which describes how likely something is to happen. The sections below show how to calculate basic probabilities using set theory, counting principles, and probability distributions.

Logic and Set Theory

Logic

Mathematical logic is a systematic method of determining the truth of a **proposition**, or statement. A proposition can be true or false, a label called the proposition's **truth value**. In the context of logical arguments, a statement cannot be both true and false; it also cannot be neither true nor false. Propositions are represented by variables, usually p, q, or r.

A **negation** has the opposite truth value of the original statement, and is often denoted by the symbol tilde (~). The statement $\sim p$ is read as "not p." Examples of statements and their negations are below. Note that $\sim p$ is not necessarily false; it is simply the opposite of p.

- p: 4 is an even number (true)

  $\sim p$: 4 is not an even number (false)
- p: Dogs lay eggs (false)

  $\sim p$: Dogs do not lay eggs (true)

A **truth table** shows all the possible inputs and truth output values of a statement or proposition. These tables are constructed by writing the variables for each statement and operation across the top row, and then listing all possible true/false values in the columns. In the table below, the statement p and its negation are in the top row, and the possible true/false values for *p* are in the left column. The right column (~*p*) can then be filled in.

p	~*p*
T	F
F	T

Figure 5.1. Truth Table

A **conjunction** between two variables or statements is an *and* statement. It is true only when both variables or statements are true. For all other situations, the result is false. A conjunction between statements *p* and *q* is written *p* ∧ *q*.

p	*q*	*p* ∧ *q*
T	T	T
T	F	F
F	T	F
F	F	F

Figure 5.2. Conjunction

A **disjunction** between two variables or statements is an inclusive *or* statement, and is denoted *p* ∨ *q*. This statement is true whenever p is true, *q* is true, or both are true.

p	*q*	*p* ∨ *q*
T	T	T
T	F	T
F	T	T
F	F	F

Figure 5.3. Disjunction

An **implication** statement is an *if... then* statement. The statement "if *p*, then *q*" is also written as *p* → *q*. The implication statement is false only when the first proposition is true and the second is false:

p	*q*	*p* → *q*
T	T	T
T	F	F
F	T	T
F	F	T

Figure 5.4. Implication

The **biconditional** or **equivalence** statement is true whenever both p and q have the same truth value:

p	q	$p \leftrightarrow q$
T	T	T
T	F	F
F	T	F
F	F	T

Figure 5.5. Bioconditional

Truth tables can be used to show that two statements are logically equivalent by showing that they have the same truth values under all circumstances. For example, a truth table can prove that the implication statement $p \rightarrow q$ and the contrapositive of that statement $\sim q \rightarrow \sim p$ are equivalent. To construct the table, create columns for p and q, and fill these with all possible true/false combinations. Next, create columns for each operation and find its truth value. Because the columns for $p \rightarrow q$ and $\sim q \rightarrow \sim p$ are identical, the statements are equivalent.

p	q	$\sim p$	$\sim q$	$p \rightarrow q$	$\sim q \rightarrow \sim p$
T	T	F	F	T	T
T	F	F	T	F	F
F	T	T	F	T	T
F	F	T	T	T	T

Figure 5.6. Truth Table with Two Statements

SAMPLE QUESTION

1) **Prove De Morgan's theorem $\sim(p \vee q) = \sim p \wedge \sim q$ using a truth table.**

Answer:

Begin by making a table with columns for p, q, $\sim p$, and $\sim q$. Then create and fill in columns for the left statement and the right statement.

p	q	$\sim p$	$\sim q$	$p \vee q$	$\sim(p \vee q)$	$\sim p \wedge \sim q$
T	T	F	F	T	F	F
T	F	F	T	T	F	F
F	T	T	F	T	F	F
F	F	T	T	F	T	T

The last two columns have the exact same truth values, which means the statements are logically equivalent. Thus, **the proposition is a true statement.**

SET THEORY

A **set** is any collection of items. In mathematics, a set is represented by a capital letter and described inside curly brackets. For example, if S is the set of all integers less than 10, then $S = \{x | x$ is an integer and $x < 10\}$. The vertical bar | is read *such that*. The set that contains no elements is called the **empty set** or the **null set** and is denoted by empty brackets { } or the symbol $\varnothing$.

Usually there is a larger set that any specific problem is based in, called the **universal set** or **U**. For example, in the set S described above, the universal set might be the set of all real numbers. The **complement** of set A, denoted by $\overline{A}$ or A', is the set of all items in the universal set, but NOT in A. It can be helpful when working with sets to represent them with a **Venn diagram**.

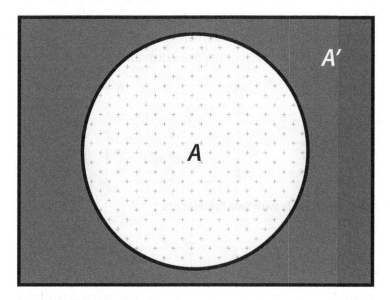

Figure 5.7. Venn Diagram

Oftentimes, the task will be working with multiple sets: A, B, C, etc. A **union** between two sets means that the data in both sets is combined into a single, larger

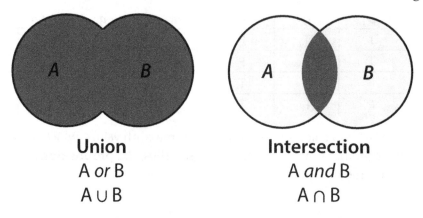

Union	Intersection
A *or* B	A *and* B
A ∪ B	A ∩ B

Figure 5.8. Unions and Intersections

set. The union of two sets, denoted $A \cup B$ contains all the data that is in either set A or set B or both (called an **inclusive or**). If $A = \{1, 4, 7\}$ and $B = \{2, 4, 5, 8\}$, then $A \cup B = \{1, 2, 4, 5, 7, 8\}$ (notice 4 is included only once). The **intersection** of two sets, denoted $A \cap B$ includes only elements that are in both A and B. Thus, $A \cap B = \{4\}$ for the sets given above.

If there is no common data in the sets in question, then the intersection is the null set. Two sets that have no elements in common (and thus have a null in the intersection set) are said to be **disjoint**. The **difference $B - A$** or **relative complement** between two sets is the set of all the values that are in B, but not in A. For the sets defined above, $B - A = \{2, 5, 8\}$ and $A - B = \{1, 7\}$. The relative complement is sometimes denoted as $B\backslash A$.

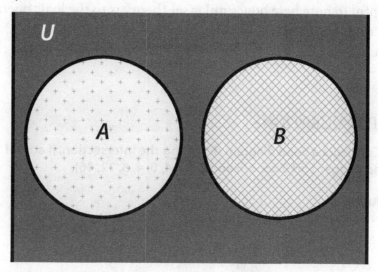

Figure 5.9. Disjoint Sets

Mathematical tasks often involve working with multiple sets. Just like numbers, sets and set operations have identities and properties.

Set Identities

$A \cup \emptyset = A$	$A \cup U = U$	$A \cup \overline{A} = U$
$A \cap \emptyset = A$	$A \cap U = A$	$A \cap \overline{A} = \emptyset$

Set Properties

Commutative Property	$A \cup B = B \cup A$	$A \cap B = B \cap A$
Associative Property	$A \cup (B \cup C) = (A \cup B) \cup C$	$A \cap (B \cap C) = (A \cap B) \cap C$
Distributive Property	$A \cup (B \cap C) = (A \cup B) \cap (A \cup C)$	$A \cap (B \cup C) = (A \cap B) \cup (A \cap C)$

De Morgan's Laws

$$\overline{(A \cup B)} = \overline{A} \cap \overline{B} \qquad \Big| \qquad \overline{(A \cap B)} = \overline{A} \cup \overline{B}$$

The number of elements in a set A is denoted $n(A)$. For the set A above, $n(A) = 3$, since there are three elements in that set. The number of elements in the union of two sets is $n(A \cup B) = n(A) + n(B) - n(A \cap B)$. Note that the number of elements in the intersection of the two sets must be subtracted because they are being counted twice, since they are both in set A and in set B. The number of elements in the complement of A is the number of elements in the universal set minus the number in set A: $n(\overline{A}) = n(U) - n(A)$.

It is helpful to note here how similar set theory is to the logic operators of the previous section: negation corresponds to complements, the "and" ($\wedge$) operator to intersection ($\cap$), and the "or" ($\vee$) operator to unions ($\cup$); notice even the symbols are similar.

SAMPLE QUESTIONS

2) Construct a Venn diagram depicting the intersection, if any, of $Y=\{x \mid x$ is an integer and $0 < x < 9\}$ and $Z = \{-4, 0, 4, 8, 12, 16\}$.

Answer:

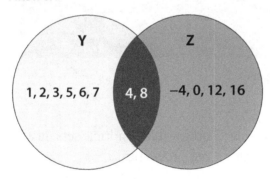

3) Suppose the universal set U is the set of all integers between -10 and 10 inclusive. If $A = \{x \in U \mid x$ is a multiple of $5\}$ and $B = \{x \in U \mid x$ is a multiple of $2\}$ are subsets within the universal set, find $\overline{A}$, $A \cup B$ and $A \cap B$, and $\overline{A} \cap \overline{B}$.

A. $\overline{A}$

B. $A \cup B$

C. $A \cap B$

D. $\overline{A} \cap \overline{B}$

Answer:

A. $\overline{A}$ includes all elements of the universal set that are not in set A:

$\overline{A} = \{-9, -8, -7, -6, -4, -3, -2, -1, 1, 2, 3, 4, 6, 7, 8, 9\}$.

B. $A \cup B$ is all elements in either A or B:

$A \cup B = \{-10, -5, 0, 5, 10, -8, -6, -4, -2, 2, 4, 6, 8\}$

C. $A \cap B$ is all elements in both A and B:

$A \cap B = \{-10, 0, 10\}$

D. $\overline{A} \cap \overline{B}$ is all the elements of the universal set that are not in either A or B:

$\overline{A} \cap \overline{B} = \{-9, -7, -3, -1, 1, 3, 7, 9\}$

PROBABILITY

Probability describes how likely something is to happen. In probability, an **event** is the single result of a trial, and an **outcome** is a possible event that results from a trial. The collection of all possible outcomes for a particular trial is called the **sample space**. For example, when rolling a die, the sample space is the numbers 1 – 6. Rolling a single number, such as 4, would be a single event.

COUNTING PRINCIPLES

Counting principles are methods used to find the number of possible outcomes for a given situation. The **fundamental counting principle** states that, for a series of independent events, the number of outcomes can be found by multiplying the number of possible outcomes for each event. For example, if a die is rolled (6 possible outcomes) and a coin is tossed (2 possible outcomes), there are $6 \times 2 = 12$ total possible outcomes.

Combinations and permutations describe how many ways a number of objects taken from a group can be arranged. The number of objects in the group is written n, and the number of objects to be arranged is represented by r (or k). In a **combination**, the order of the selections does not matter because every available slot to be filled is the same. Examples of combinations include:

Figure 5.10. Fundamental Counting Principle

▸   picking 3 people from a group of 12 to form a committee (220 possible committees)

▸   picking 3 pizza toppings from 10 options (120 possible pizzas)

In a **permutation**, the order of the selection matters, meaning each available slot is different. Examples of permutations include:

▸   handing out gold, silver, and bronze medals in a race with 100 participants (970,200 possible combinations)

▸   selecting a president, vice-president, secretary, and treasurer from among a committee of 12 people (11,880 possible combinations)

The formulas for the both calculations are similar. The only difference—the $r!$ in the denominator of a combination—accounts for redundant outcomes. Note that both permutations and combinations can be written in several different shortened notations.

$$\text{Permutation: } P(n,r) = {}_nP_r = \frac{n!}{(n-r)!}$$
$$\text{Combination: } C(n,r) = {}_nC_r = \binom{n}{r} = \frac{n!}{(n-r)!r!}$$

SAMPLE QUESTIONS

5) A personal assistant is struggling to pick a shirt, tie, and cufflink set that go together. If his client has 70 shirts, 2 ties, and 5 cufflinks, how many possible combinations does he have to consider?

Answer:

Multiply the number of outcomes for each individual event:

$(70)(2)(5) = $ **700 outfits**

6) If there are 20 applicants for 3 open positions, in how many different ways can a team of 3 be hired?

Answer:

The order of the items doesn't matter, so use the formula for combinations:

$C(n,r) = \frac{n!}{(n-r)!r!}$

$C(20,3) = \frac{20!}{(20-3)!3!}$

$= \frac{20!}{(17!\,3!)}$

$= \frac{(20)(19)(18)}{3!} = $ **1140 possible teams**

7) Calculate the number of unique permutations that can be made with five of the letters in the word *pickle*.

Answer:

To find the number of unique permutations of 5 letters in pickle, use the permutation formula:

$P(n,r) = \frac{n!}{(n-r)!}$

$P(6,5) = \frac{6!}{(6-5)!}$

$= \frac{720}{1} = $ **720**

8) Find the number of permutations that can be made out of all the letters in the word *cheese*.

Answer:

The letter *e* repeats 3 times in the word *cheese*, meaning some permutations of the 6 letters will be indistinguishable from others. The number of permutations must be divided by the number of ways the three *e*'s can be arranged to account for these redundant outcomes:

$$\text{total number of permutations} = \frac{\text{number of ways of arranging 6 letters}}{\text{number of ways of arranging 3 letters}} = \frac{6!}{3!} = 6 \times 5 \times 4 = \mathbf{120}$$

PROBABILITY OF A SINGLE EVENT

The probability of a single event occurring is the number of outcomes in which that event occurs (called **favorable events**) divided by the number of items in the sample space (total possible outcomes):

$$P\,(\text{an event}) = \frac{\text{number of favorable outcomes}}{\text{total number of possible outcomes}}$$

The probability of any event occurring will always be a fraction or decimal between 0 and 1. It may also be expressed as a percent. An event with 0 probability will never occur and an event with a probability of 1 is certain to occur. The probability of an event not occurring is referred to as that event's **complement**. The sum of an event's probability and the probability of that event's complement will always be 1.

SAMPLE QUESTIONS

9) **What is the probability that an even number results when a six-sided die is rolled? What is the probability the die lands on 5?**

   Answer:
   $$P(\text{rolling even}) = \frac{\text{number of favorable outcomes}}{\text{total number of possible outcomes}} = \frac{3}{6} = \frac{1}{2}$$
   $$P(\text{rolling 5}) = \frac{\text{number of favorable outcomes}}{\text{total number of possible outcomes}} = \frac{1}{6}$$

10) **Only 20 tickets were issued in a raffle. If someone were to buy 6 tickets, what is the probability that person would not win the raffle?**

   Answer:
   $$P(\text{not winning}) = \frac{\text{number of favorable outcomes}}{\text{total number of possible outcomes}} = \frac{14}{20} = \frac{7}{10}$$
   or
   $$P(\text{not winning}) = 1 - P(\text{winning}) = 1 - \frac{6}{20} = \frac{14}{20} = \frac{7}{10}$$

11) **A bag contains 26 tiles representing the 26 letters of the English alphabet. If 3 tiles are drawn from the bag without replacement, what is the probability that all 3 will be consonants?**

Answer:

$$P = \frac{number\ of\ favorable\ outcomes}{total\ number\ of\ possible\ outcomes}$$

$$= \frac{number\ of\ 3\text{-consonant combinations}}{number\ of\ 3\text{-tile combinations}}$$

$$= \frac{_{21}C_3}{_{26}C_3}$$

$$= \frac{1330}{2600}$$

$$= 0.511 = \mathbf{51\%}$$

PROBABILITY OF MULTIPLE EVENTS

If events are **independent**, the probability of one occurring does not affect the probability of the other event occurring. Rolling a die and getting one number does not change the probability of getting any particular number on the next roll. The number of faces has not changed, so these are independent events.

If events are **dependent**, the probability of one occurring changes the probability of the other event occurring. Drawing a card from a deck without replacing it will affect the probability of the next card drawn because the number of available cards has changed.

HELPFUL HINT

When drawing objects, the phrase *with replacement* describes independent events, and *without replacement* describes dependent events.

To find the probability that two or more independent events will occur (*A* and *B*), simply multiply the probabilities of each individual event together. To find the probability that one, the other, or both will occur (*A* or *B*), it's necessary to add their probabilities and then subtract their overlap (which prevents the same values from being counted twice).

Conditional probability is the probability of an event occurring given that another event has occurred. The notation $P(B|A)$ represents the probability that event *B* occurs, given that event *A* has already occurred (it is read "probability of *B*, given *A*").

Table 5.1. Probability Formulas

Independent Events	Intersection *and*	$P(A \cap B) = P(A) \times P(B)$	
	Union *or*	$P(A \cup B) = P(A) + P(B) - P(A \cap B)$	

Dependent Events	Conditional	$P(B|A) = P(A \cap B)/P(A)$	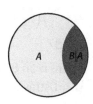

Two events that are **mutually exclusive** CANNOT happen at the same time. This is similar to disjoint sets in set theory. The probability that two mutually exclusive events will occur is zero. **Mutually inclusive** events share common outcomes.

SAMPLE QUESTIONS

12) **A card is drawn from a standard 52 card deck. What is the probability that it is either a queen or a heart?**

Answer:

This is a union (*or*) problem.

$P(A)$ = the probability of drawing a queen = $\frac{1}{13}$

$P(B)$ = the probability of drawing a heart = $\frac{1}{4}$

$P(A \cap B)$ = the probability of drawing a heart and a queen = $\frac{1}{52}$

$P(A \cup B) = P(A) + P(B) - P(A \cap B)$

$= \frac{1}{13} + \frac{1}{4} - \frac{1}{52}$

$= \mathbf{0.31}$

13) **A group of ten individuals is drawing straws from a group of 28 long straws and 2 short straws. If the straws are not replaced, what is the probability, as a percentage, that neither of the first two individuals will draw a short straw?**

Answer:

This scenario includes two events, *A* and *B*.

The probability of the first person drawing a long straw is an independent event:

$P(A) = \frac{28}{30}$

The probability the second person draws a long straw changes because one long straw has already been drawn. In other words, it is the probability of event *B* given that event *A* has already happened:

$P(B|A) = \frac{27}{29}$

The conditional probability formula can be used to determine the probability of both people drawing long straws:

$P(A \cap B) = P(A)P(B|A)$

$$= \left(\tfrac{28}{30}\right)\left(\tfrac{27}{29}\right)$$

$$= 0.87$$

There is an **87% chance** that neither of the first two individuals will draw short straws.

BINOMIAL PROBABILITY

A binomial (or Bernoulli) trial is an experiment with exactly two mutually exclusive outcomes (often labeled success and failure) where the probability of each outcome is constant. The probability of success is given as p, and the probability of failure is $q = 1 - p$. The **binomial probability** formula can be used to determine the probability of getting a certain number of successes (r) within a given number of trials (n). These values can also be used to find the expected value (μ), or mean, of the trial, and its standard deviation (σ).

$$P = {}_nC_r(p^r)(q^{n-r})$$

$$\mu = np$$

$$\sigma = \sqrt{np(1-p)}$$

SAMPLE QUESTION

14) **What is the probability of rolling a five on a standard 6-sided die 4 times in 10 tries?**

Answer:

$p = \tfrac{1}{6}$ $q = \tfrac{5}{6}$ $n = 10$ $r = 4$	Identify the variables given in the problem.
$P = {}_nC_r(p^r)(q^{n-r})$ $= \left(\tfrac{10!}{(10-4)!4!}\right)\left(\tfrac{1}{6}\right)^4\left(\tfrac{5}{6}\right)^{10-4}$ $= 0.054$ There is a **5.4% chance**.	Plug these values into the binomial probability formula.

PROBABILITY DISTRIBUTIONS AND EXPECTED VALUE

RANDOM VARIABLES

A **random variable** is a variable whose value depends on a random event. Random variables are usually denoted with a capital X or Y. A random variable may be discrete or continuous. Discrete random variables have a finite number of specific possible outcomes. An example of a discrete random variable would be how many "heads" outcomes result when a coin is flipped three times. The distinct possible values for this variable would be 0, 1, 2, or 3, as either zero, one, two, or three heads would result upon three flips of a coin.

A **continuous** random value is defined over an interval of values, and has infinitely many possible values. An example of a continuous random variable is the amount of time an airplane is delayed. There are many, many possible outcomes in this situation. Usually, if something is being measured (time, weight, height, etc.), the variable is continuous.

Once a random variable is assigned, it's possible to find the probability for each of the that variable's values. In the example of flipping a coin three times, let X be the random variable that represents how many "heads" outcomes result. Note that each of the four possible values for this random variable are NOT equally likely, even though the possibility of tossing a head or a tail is equally likely. The sample space of the experiment of flipping a coin three times is {HHH, HHT, HTH, HTT, THH, THT, TTH, TTT}. The probabilities of random variable X taking each of its possible values can be displayed in a table. This is the **probability distribution** for each value of X:

X:	0	1	2	3
$P(X)$:	0.125	0.375	0.375	0.125

This distribution also can be displayed graphically:

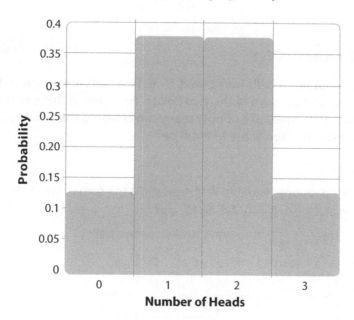

Figure 5.11. Probability Distribution

Notice that the sum of the probabilities is 1, as is the sum of the areas of the rectangles of the histogram. The sum of the probabilities of any random variable's output is always 1. Essentially, defining a random variable defines a function $P(X)$, the output of which are probabilities (making the range $0 \leq X \leq 1$), as a function of the random variable X; this is a **probability distribution function**.

Other probabilities also can be calculated using the above table. For example, to find the probability that either two or three heads occur, $P(X = 2 \text{ or } X = 3) = P(X = 2) + P(X = 3) = 0.375 + 0.125 = 0.5$. The probability that at least one head occurs is $1 - P(X = 0) = 1 - 0.125 = 0.875$ (using the complement rule).

HELPFUL HINT

Expected value can be used to weigh the possible outcomes of a decision by assigning probabilities to payoff values and determining which expected value is most beneficial.

For continuous random variables, the total area under the probability distribution curve will also be 1. Since the number of possible values taken on by a continuous random variable is infinite, the probability of any single value of X is 0. Instead, probabilities of intervals are calculated. These probabilities can be calculated by finding the area under the distribution curve (the integral) over that interval.

The **expected value**, $E(x)$, of a discrete random variable is the weighted average (or mean) of the variable. To calculate the expected value, calculate the sum of the products of each data point multiplied by the probability of that point. That is, for data points $x_1, x_2, x_3, \dots x_n$,

$$E(X) = \sum_{i=1}^{n} x_i P(x_i)$$

SAMPLE QUESTIONS

15) **A bag contains 6 balls numbered 1 – 6. Two balls are removed from the bag, and the sum of the two balls is recorded. If this experiment is repeated 50 times, about how many times would the sum be 3? What would the average value of a roll be?**

Answer:

Let X be a random variable that represents the sum of two balls. The possible values of X are 3, 4, 5, 6, 7, 8, 9, 10, and 11.

There are ${}_6C_2 = \frac{6!}{4!2!} = 15$ ways to choose two ball:

	1	2	3	4	5	6
1						
2	3					
3	4	5				
4	5	6	7			
5	6	7	8	9		
6	7	8	9	10	11	

The frequency of each *X* value can be found by counting how many times it appears in the table. The probability of each *X* value will then be the frequency divided by the total number of outcomes (15).

X (sum)	Frequency	P(X)
3	1	$\frac{1}{15}$
4	1	$\frac{1}{15}$
5	2	$\frac{2}{15}$
6	2	$\frac{2}{15}$
7	3	$\frac{1}{5}$
8	2	$\frac{2}{15}$
9	2	$\frac{2}{15}$
10	1	$\frac{1}{15}$
11	1	$\frac{1}{15}$

To find the number of times 3 would appear in 50 trials, use the expected value equation with the single probability:

$$\mu = np$$
$$50\left(\frac{1}{15}\right) = 3.33 \text{ times}$$

To find the average value, use the expected value equation:

$$E(X) = \sum_{(i=1)}^{9} x_i\, P(x_i) = 3\left(\frac{1}{15}\right) + 4\left(\frac{1}{15}\right) + 5\left(\frac{2}{15}\right) + \dots + 11\left(\frac{1}{15}\right)$$
$$= \mathbf{7}$$

16) **Suppose data is collected on the length of delays in the air travel industry. The following data are collected (data in minutes):**

5	5	10	10	11	11	18	28	25	23	32	33	34
35	36	36	36	39	45	55	57	72	74	88	90	94

Determine the probability that if a customer experiences a delay, it will be less than 40 minutes.

Answer:

Begin by defining random variable *X*, which represents the amount of time of a delay. Organize the data into intervals:

Delay times X:	$0 < X \le 20$	$20 < X \le 40$	$40 < X \le 60$	$60 < X \le 80$	$80 < X \le 100$
P(a < X ≤ b):	0.269	0.423	0.115	0.077	0.115

The probability the delay is less than 40 minutes is:

$P(0 < X \le 20 \text{ or } 20 < X \le 40) = 0.26 + 0.423 = \mathbf{0.692}$

THE NORMAL DISTRIBUTION

The **normal distribution** or **normal curve** is an example of a continuous probability distribution. The normal curve is a bell-shaped curve that occurs frequently in natural phenomena. Normal distributions are symmetric about the mean of the data. For a perfectly symmetric normal curve, the median and the mean are the same. The data that fit a normal curve are distributed as follows:

▸ Approximately 68% of the data lies within 1 standard deviation of the mean (between $\bar{x} - s$ and $\bar{x} + s$).

▸ Approximately 95% of the data lies within 2 standard deviations of the mean (between $\bar{x} - 2s$ and between $\bar{x} + 2s$).

▸ Approximately 99.7% of the data lies within 2 standard deviations of the mean (between $\bar{x} - 3s$ and between $\bar{x} + 3s$).

The shape of the curve varies somewhat based on the magnitude of the standard deviation. If the deviation is small, the curve will be horizontally compressed and will be tall and thin. If the deviation is large, the curve will be horizontally stretched and will appear flatter.

The area under the normal curve is 1. The probability of any given interval can be found by calculating the area under the curve over that interval. A graphing calculator may be used to calculate areas on intervals.

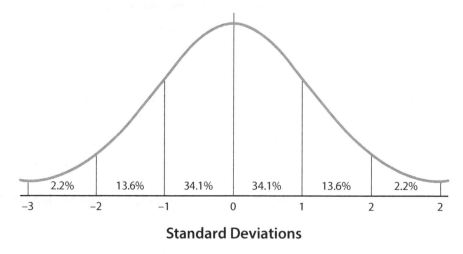

Figure 5.12. Normal Distribution

SAMPLE QUESTION

17) A common example of a normal curve distribution is the IQ test, as shown below. What is the probability that a person chosen at random has an IQ lower than 130?

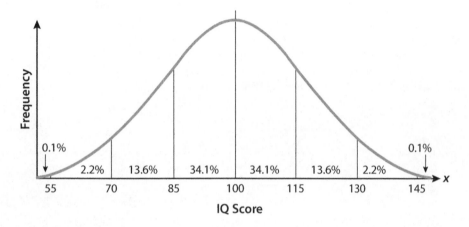

Answer:

The probability that a person chosen at random has an IQ lower than 130 is $1 - P$(IQ is above 130)

$= 1 - (0.022 + 0.001) = $ **0.977, or 97.7%.**

6

Curriculum, Instruction, and Assessment

CURRICULUM OUTCOMES AND STUDENT LEARNING OBJECTIVES

CURRICULUM MAP

A successful mathematics course begins with a **curriculum map**. A curriculum map is a long-term plan that lays out the content to be covered during the course. A curriculum map is not as detailed as a lesson or unit plan; however, the curriculum map does break the academic year into specific units outlining when students will be expected to master necessary learning outcomes. It is imperative that the map give an estimation of the time needed to cover each unit, in order to ensure that all of the necessary content is covered before the end of the course. The timeline also serves as a resource when planning each individual unit—having a finite, set number of days to cover a particular concept will make it less tempting to cover unnecessary material.

A curriculum map shows the learning outcomes, or standards, to be covered during each unit. The map can also break each unit into the concepts and procedural skills to be learned, as well as any necessary vocabulary that the students need to master. In addition, the curriculum map can show any mathematical practices necessary for successful completion of the unit. Incorporating practices such as abstract reasoning, constructing arguments, critiquing the reasoning of others, and creating mathematical models is necessary in order for students to develop a deeper, conceptual understanding of the course content. The curriculum map will be used to create unit plans in which the teacher will write measurable learning objectives based on the outcomes that the students need to master.

LEARNING OUTCOMES

Learning outcomes, also called **standards**, are general statements that summarize content to be mastered upon completion of a unit or course. The outcomes describe high-level skills that the student is expected to master. High-level skills include applying the content in order to successfully analyze, evaluate, and create. The outcomes are measurable by evaluation or observation. Often a rubric is used as a tool to assist in measuring the learning outcomes.

An example of an Algebra 1 learning outcome is as follows: *Understand the relationship between the zeros and the factors of a polynomial.*

LEARNING OBJECTIVES

Student learning objectives are the individual pieces that guide the student to mastery of each learning outcome. All levels of Bloom's Taxonomy should be included in the objectives in order for the students to reach mastery of the content. Mastery of the content can be measured with assessments and activities that ask the students to apply higher-order thinking; the students should be asked to apply the content learned in order to analyze, evaluate, and create. Like the learning outcomes, the learning objectives should be measurable against some form of criteria. Tracking each student's progress serves as an aid in identifying misconceptions and areas that need reinforcement. Perhaps a student is struggling with a particular concept, or perhaps the majority of the students have failed to master necessary objectives, and the material needs to be covered differently. By using the learning objectives as a guide throughout the unit, the teacher can provide specific and meaningful feedback to each student to promote mastery of the learning outcomes.

Before a student can master the previously stated Algebra 1 learning outcome—*understand the relationship between the zeros and the factors of a polynomial*—the student must understand other concepts ranging from concrete to more abstract. It would be appropriate to write the following list of learning objectives based on this learning outcome:

1. Understand the definition of a polynomial and identify polynomials.
2. Write a polynomial in standard form.
3. Determine the degree of a polynomial.
4. Determine the leading coefficient of a polynomial.
5. Add and subtract polynomials.
6. Multiply polynomials.
7. Factor trinomials with a leading coefficient of one.
8. Factor trinomials with a leading coefficient other than one.
9. Factor trinomials with a common factor.
10. State the relationship between the factored form of a polynomial and the corresponding graph.

11. Determine the relationship between the zeros of a polynomial and the factored form of the polynomial.

The objectives for this particular learning outcome begin with the more basic and concrete skills. Understanding the definition of and identifying a polynomial, and adding, subtracting, multiplying, and factoring polynomials are the basic skills students need to master before they are able to master the learning outcome. These skills range from basic to more difficult, and should be taught as such. Each unit might only have three or four outcomes but many objectives throughout the unit scaffold the content and lead the students to content mastery. It is important to give students all of the necessary skills and set them up for mastery of a more abstract concept. As seen in the learning objectives above, learning objective 10 begins leading the students closer to the more abstract and higher-level learning outcome.

In order for the students to gain an understanding of the more abstract ideas, the teacher cannot simply state: if $x - a$ is a factor of a polynomial, then a is a zero of the polynomial. Likewise, the teacher cannot simply state: if a is a zero of a polynomial, then $x - a$ is a factor. The teacher should scaffold the lessons in such a way that the students ultimately arrive at these ideas.

For instance, creating a lesson plan that revolves around objective 10—*state the relationship between the factored form of a polynomial and the corresponding graph*—will help the students begin to visualize the relationship between factored polynomials and the equivalent graph. If the students have already learned how to graph a quadratic function, this lesson can also serve as a review for that particular skill. If the students have not learned how to graph a quadratic function, a free online graphing program, such as Desmos, or an appropriate graphing calculator can be utilized. If every student does not have access to a computer or device necessary to access online software, the lesson can be done in groups or as a class. If the activity is done as a class, the teacher should ensure that students are taking notes and are actively involved and participating in the lesson.

Regardless of the specifics of the lesson plan, it should be robust in the skills and ideas students are asked to investigate. It is much more meaningful to allow the students to perform this investigation so that they have the opportunity to make the connection between the relationship of a factored polynomial and its graph. If this lesson is being done together as a class, it might be helpful to project the graph on a screen and allow the students to perform the analysis. By using a quick assessment at the end of class, the teacher can gauge student understanding of the more abstract ideas. If the assessment shows that the students have failed to make any connection, perhaps another lesson structured in a different manner is necessary. Collaborative work during the next lesson may be a helpful way for students to work through the material, allowing students to speak to one another and debate their understandings regarding the material. This provides the students who have mastered the content an opportunity to verbalize and explain their understanding, while allowing the students who have not mastered the content an opportunity to learn from classmates.

Once students are able to express an understanding of the connection between the factored form of a polynomial and the corresponding graph, the students can begin to determine the relationship between the zeros of a polynomial and the factored form of the polynomial. Each tool that the students need in order to master the ultimate goal is carefully and thoughtfully arranged in the unit plan. It is important to note that perhaps ultimate mastery of the goal cannot be achieved until additional units are covered.

For instance, consider the polynomial $x^2 - 3x - 4 = 0$. Also consider the development of the concept: if $x - a$ is a factor of a polynomial, then a is a zero of the polynomial. Through the careful scaffolding of learning objectives, students should have learned the skill of factoring and be able to determine that the factors of the given polynomial are $(x - 4)$ and $(x + 1)$. Setting each factor equal to zero and solving for x results in two answers: $x = 4$ and $x = -1$. Note that this is an opportunity to determine whether the students are stating that the answers are $x = -4$ and $x = 1$. This is a common mistake and can be corrected by requiring students to show each step of the process: $x - 4 = 0$; add 4 to each side; $x = 4$. By graphing the polynomial and analyzing the graph, either by hand or using appropriate technology, students can easily see that the graph crosses the x-axis at -1 and 4.

After analyzing a few problems of this type, students will begin to show an understanding of the relationship between the zeros and factored form. Perhaps interpreting the meaning of these values in a real-world scenario is a separate learning outcome. However, each tool that the students develop deepens and strengthens their understanding of polynomials. If analyzing real-world projectile scenarios is to be covered in a different unit, it might be appropriate to introduce a few basic situations at this time. This could be the key for some students to put the final pieces of the puzzle together regarding the learning outcome we are concerned with. It could also provide variety for students who are quickly mastering the content. This may seem confusing to some students who are struggling with the content, but this concept will be covered again in a future unit.

This situation shows that not only are the learning objectives connected to one another and to the learning outcomes, but that the outcomes are also connected to one another. These are all things to consider when creating a year-long plan, as well as individual unit and lesson plans.

SAMPLE QUESTIONS

1) **Which of the following statements accurately describes learning outcomes and objectives?**

   A. Learning outcomes summarize content to be mastered upon completion of a unit, while learning objectives summarize content to be mastered upon completion of a course.

   B. Learning outcomes typically describe high-level skills, while learning objectives describe skills ranging anywhere from basic to abstract.

   C. Learning outcomes form the basis of each lesson, while learning objectives summarize high-level content to be mastered upon completion of a unit or course.

   D. Learning outcomes describe skills and procedural knowledge, while learning objectives describe the ability to analyze, evaluate, and create.

   Answers:

   A. Incorrect. Learning *outcomes* summarize content to be mastered upon completion of a unit or a course.

   B. **Correct.** Learning outcomes test a student's ability to apply knowledge in order to analyze, evaluate, and create; objectives help the student build knowledge ranging from basic skills and more abstract concepts.

   C. Incorrect. Learning *objectives* form the basis of each lesson while learning *outcomes* summarize high-level content to be mastered upon completion of a unit or course.

   D. Incorrect. Learning *outcomes* describe the ability to analyze, evaluate, and create while learning *objectives* describe skills, procedural knowledge, and the ability to apply knowledge to higher order concepts.

2) **Which of the following statements best describes a method for correcting a recurring mistake?**

   A. having the student label parts of the problem and show all work

   B. allowing the student to use a calculator

   C. allowing the student to choose when to mentally solve for necessary values

   D. allowing the student to show work for only the most difficult parts of a problem

   Answers:

   A. **Correct.** When a student repeatedly makes a mistake, asking the student to label the parts of a problem before beginning and show all of the steps of the problem-solving process will decrease the chances of making a careless error.

B.   Incorrect. A calculator can be a great tool, but the calculator will only give you the answer to whatever problem you enter. Entering the wrong information will still result in a wrong answer.

C.   Incorrect. In order to correct a mistake that continues to happen, students need to show all work rather than pick and choose when to show work.

D.   Incorrect. Again, in order to correct a recurring mistake, it is best to have students show all work rather than pick and choose when to show work.

COMPONENTS OF A GOALS-BASED UNIT

UNIT PLAN

Prior to beginning a unit of instruction in a mathematics course, it is essential to create a **unit plan**. The overall unit plan serves as a road map for a particular unit. Each unit outlined in the curriculum map should have a detailed unit plan prior to implementation. The unit plan is made of appropriate and relevant learning outcomes, student learning objectives, a pre-unit assessment, a summative unit assessment, and specific lesson plans. Not only does creating a unit plan aid in organization and preparation for the teaching of the unit, it also helps the teacher foresee difficulties and misconceptions that the students may already have or encounter during the unit.

In order to create a successful unit of study in mathematics, it is necessary to consider the exact skills and concepts the students should be able to do or know by the end of the unit. What does successful student mastery of the learning outcomes look like? The summative unit assessment will test for mastery of the learning outcomes, so it is necessary to have a clear vision of what the students will have learned as well as a clear understanding of how the students will show their mastery of the content.

CREATING A UNIT PLAN

One of the most effective types of unit planning is called **backward unit design**. Backward unit design involves first developing a clear and precise understanding of what student achievement of the learning outcomes looks like. How will students demonstrate mastery of the skills and concepts covered in the unit? After developing this understanding, the teacher will design a **summative unit assessment** that measures student mastery of the unit outcomes. By designing the unit assessment first, the teacher can make more informed decisions about exactly what lessons will be most effective at teaching the content at the appropriate level of rigor. See *Assessment* below for more details.

By utilizing the backward unit design method, teachers can ensure that both the method used to deliver the content as well as the method used to measure student mastery is rigorous enough. When teachers test what they teach—rather than develop the assessment first—they run the risk of failing to actually meet the learning outcomes. Developing content before assessment leads to lower levels of student achievement. For instance, a teacher may develop a six-week unit of lessons aimed at achieving certain learning outcomes *without* considering what mastery of these outcomes actually looks like. During week five, the teacher develops the summative unit assessment based on the skills and knowledge taught to the students during the unit. The unit assessment could accurately measure student mastery of the skills and concepts covered during the unit. But what if the teacher has failed to teach the content at the level of rigor necessary to develop student mastery of the unit's learning outcomes? Those students have a lower chance of performing well on an end-of-year assessment compared to students who have been prepared at the level of rigor necessary to master the learning outcomes for the course.

Remember that a summative unit assessment should measure mastery of the unit's learning outcomes rather than each individual learning objective. Hundreds of learning objectives should be covered throughout the year in order for students to master the learning outcomes. It would be impossible to formally assess each objective; however, assessment of the objectives will be done throughout each unit through the use of various assessment methods. The overall learning outcomes usually describe a deeper level of understanding of a concept, as opposed to simple skills and memorization tasks. Measuring student mastery of the learning outcomes means measuring rigorous and higher-level understandings. This makes it important to design, scaffold, and implement lessons in a way that allows students to develop deep understandings and higher-order thinking skills rather than surface-level skills and understandings. Backward unit design—rather than designing an assessment after the lessons have been implemented—helps to ensure that the lessons are rigorous enough to develop the higher-level understanding necessary for students to demonstrate mastery of the learning outcomes.

WRITING LEARNING OBJECTIVES

Now that the unit plan has learning outcomes and a summative unit assessment designed to measure mastery of each outcome, the teacher can begin to write measurable learning objectives. These learning objectives will provide the basis for each individual lesson in the unit. The atom is the basic building block of matter; likewise, the objective is the basic building block of a lesson. By converting each learning outcome into measurable learning objectives, the teacher can design and scaffold the unit's lessons in such a way that gives students the opportunity to investigate and develop rigorous skills and knowledge. There are many criteria for writing effective lesson objectives. It is important that the lesson objectives for mathematics courses are student-centered, measurable, and offer an appropriate level of rigor.

Consider learning objectives 1 – 6 written in the previous section *Learning Objectives*. Analyze the objectives and identify possible ways to improve each objective. Keep in mind that objectives should be student-centered, measurable, and offer an appropriate level of rigor.

Table 6.1. Measurable Learning Objectives

	Objective	Notes	Updated Objective
1.	Understand the definition of a polynomial and identify polynomials.	It is difficult to measure whether or not a student *understands*. This objective is not written as a student-centered statement.	The student will be able to identify and define the parts of a polynomial.
2.	Write a polynomial in standard form.	This objective is not written as a student-centered statement.	The student will be able to write a polynomial in standard form.
3.	Determine the degree of a polynomial.	This objective is not written as a student-centered statement.	The student will be able to identify and state the degree of a polynomial written in both standard and non-standard form.
4.	Determine the leading coefficient of a polynomial.	This objective is not written as a student-centered statement.	The student will be able to identify and state the leading coefficient of a polynomial.
5.	Add and subtract polynomials.	This objective is not written as a student-centered statement. At this point in the unit, students will begin to apply their knowledge of the concepts learned thus far. Learning objectives should become more specific and more rigorous.	The student will be able to apply knowledge of polynomials in order to add and subtract polynomials of the same and varying degrees. In addition, the student will be able to add and subtract polynomials with the same number of terms and different numbers of terms.
6.	Multiply polynomials.	This objective is not written as a student-centered statement. Again, students will apply their knowledge of the concepts learned thus far. Learning objectives should become more specific and more rigorous.	The student will be able to apply the distributive property to multiply polynomials with the same number of terms and with different numbers of terms. The student will be able to write the product in simplest form by combining like terms.

Notice that all of the updated objectives are student-centered, measurable, and offer an appropriate level of rigor. In addition, the objectives are written in such a way that scaffolds the skills and concepts; therefore, the objectives are more rigorous as the unit continues. This allows students to develop an understanding of the concrete skills at the beginning of the unit and apply those skills to more abstract concepts toward the end of the unit.

Pay close attention to learning objectives 5 and 6. Objective 5 states that students will be able to add and subtract polynomials. It is important to also note in the updated objective that the students will be able to add and subtract polynomials of the same degree and also add and subtract polynomials that have different degrees. Students will also be expected to understand how to add and subtract polynomials that have the same number of terms as well as different numbers of terms. Otherwise, students may develop the misconception that to add and subtract polynomials, one must simply line up the terms and perform the arithmetic. Objective 6 states that students will be able to multiply polynomials. Without taking the time to write a learning objective, the teacher may be tempted to simply teach FOIL. This is a very surface level understanding of polynomial multiplication. The updated objective—*the student will be able to apply the distributive property to multiply polynomials with the same number of terms and with different numbers of terms. The student will be able to write the product in simplest form by combining like terms*—makes sure that the teacher must cover polynomial multiplication with more rigor than simply teaching FOIL. The updated objective also provides an opportunity to review the distributive property and correct any misunderstandings students may have about the property.

Taking the time to write learning objectives will help the teacher foresee concepts that may leave room for misconception or misunderstanding. Remember that these objectives form the basis of the lessons; therefore, the more specific and rigorous the objectives, the more specific and rigorous the lesson must be to successfully teach the material.

SAMPLE QUESTION

3) **Choose the most well-written learning objective from the following list.**

   A. Determine the *x*-intercept and *y*-intercept of an equation.

   B. The student will be able to determine the *x*-intercept and *y*-intercept of an equation.

   C. Determine the *x*-intercept and *y*-intercept of an equation by solving for each intercept as well as by analyzing a graph of the equation.

   D. The student will be able to determine the *x*-intercept and *y*-intercept of an equation by solving for each intercept as well as by analyzing a graph of the equation.

Answers:

A. Incorrect. Learning objectives should be student-centered, measurable, and rigorous; learning objective A is not student-centered or specific enough.

B. Incorrect. Learning objectives should be student-centered, measurable, and rigorous; learning objective B could be written more clearly in order to measure specific skills—solving for each intercept and analyzing a graph.

C. Incorrect. Learning objectives should be student-centered, measurable, and rigorous; learning objective C is not student-centered.

D. Correct. Learning objective D is student-centered, measurable, rigorous, and identifies specific skills to be measured.

LESSON PLANNING AND MATHEMATICAL INSTRUCTION

The unit plan is not complete without specific **lesson plans** for each lesson that the teacher will teach during the unit. In order to promote problem-solving skills and allow students to investigate new mathematical concepts, the teacher should avoid lessons that rely solely on direct instruction. Instead, the teacher should scaffold each lesson in such a way that students are given the opportunity to analyze the new material and attempt to form meaningful connections between the material and previous concepts learned. Giving students time to collaborate provides a great opportunity to learn how to communicate about mathematical concepts and to critique the reasoning of others. It is imperative that students learn how to provide mathematical justification as well as analyze a given justification. It is not enough to simply state the answer; in mathematics, students must provide sound reasoning for why they believe the answer is correct. Effective lessons make use of a variety of instructional strategies in order to provide all students with opportunities to develop and improve mathematical skills and procedures. Every lesson cannot rely on the same strategy if the teacher hopes to appeal to all types of learners in the classroom.

CREATING A LESSON PLAN

INTRODUCTION

Use the first brief portion of class to introduce the new lesson. If there is a particular skill that is important for the students to review prior to beginning the lesson, use this time to give a bell ringer. The bell ringer is a set of problems that gives the students an opportunity to review the necessary skills.

Other ideas include showing a brief video, posing an interesting problem that the students will learn how to solve during the lesson, or holding a class discussion about the relevance of certain math skills in their lives.

Questions to consider:

▶ What is the learning objective?

▶ What skills do the students need for this lesson?

▶ What skills should the students gain from this lesson?

▶ What higher-order skills should the students utilize during this lesson?

▶ What real-world scenarios can be used to show relevance? Should the teacher give these examples or design the lesson so that the students have the opportunity to brainstorm these examples?

INVESTIGATION

If the lesson involves a basic procedural skill, perhaps it is most beneficial for the teacher to provide a brief period of direct instruction. A common instructional strategy for implementing direct instruction is the *I do, we do, you do* strategy. The teacher begins by showing the students a detailed list of steps necessary to solve a problem. The teacher then asks the students to work similar problems as a class. Finally, the students are given the opportunity to work problems independently.

More rigorous lessons could involve posing a question or problem to students at the beginning of the investigation portion of the lesson. Students are then given time to analyze the problem and brainstorm possible solutions. Next, students are grouped appropriately and given the opportunity to discuss, debate, and critique the reasoning of classmates. Perhaps coming together as a whole class to discuss the solution is necessary so that all students have the opportunity to understand the correct solution and the appropriate methods utilized to arrive at the solution.

It is possible that technology could enhance student understanding. Graphing calculators, iPads or other tablet devices, and computers are great resources to use to enhance learning. If technology is to be used, it is extremely important for the teacher to determine whether or not it will enhance understanding or cause confusion. Using the resource should not be more complicated than the content being learned.

The following are examples of resources that can enhance learning when used appropriately. Most of these resources also have an online or tablet application version:

1. Base ten blocks
2. Calculator/graphing calculator
3. Clock/stopwatch
4. Graph paper
5. Protractor (There are applications which allow the measurement of angles through the use of a device's camera.)
6. Ruler

During collaborative work, it is the teacher's duty to walk around and facilitate student discussion. Teachers should encourage students to *ask three before me*. This means that when a student has a question, the student should attempt to ask three classmates before asking the teacher. This encourages student-centered learning rather than teacher-centered learning. It is not necessary for the teacher to be the only individual who answers questions. Instead, the teacher should serve as a facilitator working with the students rather than being the center of the lesson.

During this time the teacher also has the opportunity to offer one-on-one attention to struggling students. If there are multiple students who need the teacher's attention, the teacher can differentiate instruction by leading a group more closely.

Questions to consider:

▶ Does the teacher need to provide direct instruction?

▶ What problem will the students focus on during this lesson?

▶ Will the students work individually, in pairs, or groups for all or part of this lesson?

▶ What technology could enhance student learning?

▶ What is the role of the teacher during this lesson?

▶ What problems does the teacher foresee the students having with the skills or concepts in this lesson?

WRAP-UP

Now that students have had an opportunity to either independently practice a skill or participate in higher order thinking, it may be necessary to include certain skills and concepts in a teacher-led discussion. During this time, the teacher will ask guiding questions in order for the students to form connections and begin to see general rules and relationships based on the lesson's content.

As a group, the class can discuss general rules or procedures that should be written based on the content of the lesson. It is very possible that individuals or groups solved a problem correctly but did not utilize the method intended by the teacher. It is to be expected that students learn and problem solve differently. Prior to implementing the lesson, the teacher should consider the various ways that students may attempt the lesson and be prepared to explain why either the various methods are acceptable or why one method is more appropriate.

Questions to consider:

▶ What skills and concepts need to be discussed in more detail during this time?

▶ What general rules or procedures can be written based on this lesson?

ASSESS

It is important to note that there are many appropriate ways that the teacher can assess and evaluate student understanding.

Throughout the introduction and investigation portions of the lesson, the teacher can assess students by actively listening to group conversations. This will allow the teacher to facilitate discussions and ask guiding questions in groups that are experiencing difficulty. Prior to implementation of the lesson, the teacher should prepare by making a list of key questions and comments that will identify student understanding.

Another method of assessment is to ask the students to write their ideas down as they work or discuss. The teacher can collect the work and informally review it to determine student progress as well as possible misconceptions occurring.

Exit tickets are a one problem assessment intended to be completed in the last few minutes of class. Teachers can also set up red, yellow, and green folders for students to self-assess. Students who place their exit ticket in the green folder feel that they have mastered the skill or concept, yellow are somewhat unsure, and red know that they need more instruction/practice.

Questions to consider:

▸ Is it necessary to give an assessment or evaluation at this time?

▸ What methods of assessment could be used during the investigation segment of the lesson, rather than waiting until the end of the lesson?

▸ What key questions or comments should the teacher listen for in order to assess student progress throughout the lesson?

ASSESSMENT

CREATING A PRE-UNIT ASSESSMENT

Creating a summative unit assessment was discussed in the previous section *Creating a Unit Plan*. To review, summative unit assessments measure and ensure student mastery of the unit outcomes. For a successful unit, it is also necessary to design a **pre-unit assessment**. The pre-unit assessment is a diagnostic tool used to determine where the teacher should begin instruction. The pre-unit assessment should include items that test student knowledge of prerequisite skills. The assessment should also include items that test skills and conceptual knowledge of the material to be taught throughout the unit. This assessment gives the teacher a strong idea of how to differentiate for various students as well as an idea of the problems and misconceptions that need to be corrected before students will experience success with the content. In addition, the assessment provides a standard to measure student performance during the unit. The teacher should ask the following questions when designing the pre-unit assessment:

1. What skills should the students already know in order to begin this unit?
2. What items will test student knowledge of prerequisite skills?
3. What skills and concepts from this unit are students likely to already know?
4. What items will test student knowledge of these skills and concepts?
5. How will the assessment be scored?

Note: A rubric that offers an explanation along with the various point values that can be earned on each question is a great way to grade assessments. Using a rubric helps the teacher remain unbiased and fair during grading, and it helps the students understand how point values were assigned.

ASSESSMENT TYPES

A **summative assessment** is an assessment intended to evaluate student learning at the end of a unit, semester, or year. Summative assessments are high stakes and are often associated with high point values. **Formative assessments**, however, are low stakes and are often associated with low or no points. Formative assessments are intended to help students identify areas of confusion and misunderstanding. Formative assessments also guide teacher instruction and allow teachers to correct misunderstandings and misconceptions quickly rather than waiting until the end of the unit. Typically, formative assessments are the assessments given throughout the unit in order to evaluate mastery of the learning objectives. The summative assessment is then given at the end of the unit in order to evaluate mastery of the overall learning outcomes.

Formal assessments are tests that have data and statistics to describe student performance. Standardized tests are the most common form of formal assessment. Formal assessment scores are compared in order to measure student performance.

Informal assessments are non-standardized tests. The majority of the assessments given throughout the academic year are informal assessments. Teachers may use rubrics or other criteria to measure individual student performance; however, student scores on informal assessments are not compared to one another to gather data and statistics.

FORMATIVE ASSESSMENT TECHNIQUES

The following table lists ideas for various formative, informal techniques to use throughout a unit of study.

Table 6.2. Assessment Techniques

Name of Technique	Description
Application Card	Each student must write down at least one real-world application for the material learned in the lesson.
Brainstorming	Ask students to write down their ideas about a certain topic or skill. Collect the ideas to determine which students are already familiar with the content.
Create a Question	Based on the content learned in the lesson, each student should write his or her own unique problem. The problems could be traded and used as a bell ringer at the beginning of the next class, or the teacher could collect the questions and choose the appropriate problems to use on an assessment.
Critiquing	After students have time to work individually on a task, they are then assigned a partner and asked to share their reasoning and critique the reasoning of their partner. This helps students learn how to verbalize ideas about mathematics as well as how to offer justification for solutions.
Exit Ticket	Exit tickets are a one problem assessment intended to be completed in the last few minutes of class. Students must turn in their ticket before they leave the room.
Free Response Question	Students are given a higher order and more conceptual question and asked to offer a solution along with a written explanation to justify the solution.
Gallery Walk	A minimum of five questions about the current math content are posted around the classroom. In groups, students travel around the room and are given set amounts of time to spend discussing each question.
Graphic Organizer	Students work either individually or in pairs/groups to create a visual representation of the ideas being covered. The teacher can provide a specific graphic organizer or allow students to freely create their own representation.
Green, Yellow, Red	Teachers can set up red, yellow, and green folders for students to self-assess. Students who place their solution in the green folder feel that they have mastered the skill or concept, yellow are somewhat unsure, and red know that they need more instruction/practice.
Quiz	Brief assessment made of fill in the blank, multiple choice, or short open-ended practice problems.

Practice Test

Work the problem, and choose the most correct answer.

1

Simplify: $(5 + 2i)(3 + 4i)$

A.   7

B.   23

C. $7 + 26i$

D. $23 + 26i$

2

If $16^{x+10} = 8^{3x}$, what is the value of x?

A.   2

B.   4

C.   5

D.   8

3

The coordinates of point A are $(7, 12)$ and the coordinates of point C are $(-3, 10)$. If C is the midpoint of $\overline{AB}$, what are the coordinates of point B?

A. $(2, 11)$

B. $(-13, 8)$

C. $(17, 14)$

D. $(-13, 11)$

4

The pie graph below shows how a state's government plans to spend its annual budget of $3 billion. How much more money does the state plan to spend on infrastructure than education?

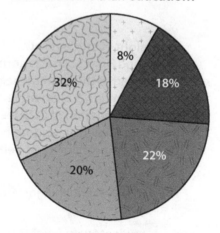

 Employees   Education

Healthcare   Pension

 Infrastructure

A. $60,000,000

B. $120,000,000

C. $300,000,000

D. $600,000,000

5

The average speed of cars on a highway (*s*) is inversely proportional to the number of cars on the road (*n*). If a car drives at 65 mph when there are 250 cars on the road, how fast will a car drive when there are 325 cars on the road?

A.   50 mph

B.   55 mph

C.   60 mph

D.   85 mph

6

Which of the following represents a linear equation?

A. $\sqrt[3]{y} = x$

B. $\sqrt[3]{x} = y$

C. $\sqrt[3]{y} = x^2$

D. $y = \sqrt[3]{x^3}$

7

The height of students at Glenwood High School is normally distributed. If the average height of a student is 5.5 feet with a standard deviation of 0.25 feet, a student picked at random has the highest probability of being in which group?

A.   Students with heights from 5 to 5.25 feet.

B.   Students with heights from 5.25 to 5.5 feet.

C.   Students with heights from 5.75 to 6 feet.

D.   Students with heights from 6 to 6.25 feet.

8

Line *a* and line *b* are perpendicular and intersect at the point $(-100, 100)$. If $(-95, 115)$ is a point on line *b*, which of the following could be a point on line *a*?

A. $(104, 168)$

B. $(-95, 115)$

C. $(-112, 104)$

D. $(-112, -104)$

9

Which of the following are NOT typically included in a well-written and rigorous learning outcome?

A.   The ability to analyze new information based on the content learned.

B.   The ability to evaluate new information based on the content learned.

C.   The ability to create a product based on the content learned.

D.   The ability to memorize information based on the content learned.

10

What is the relationship between the mean and the median in a data set that is skewed right?

A.   The mean is greater than the median.

B.   The mean is less than the median.

C.   The mean and median are equal.

D.   The mean may be greater than, less than, or equal to the median.

11

Which of the following is NOT included in a curriculum map?

A. unit titles

B. learning outcomes

C. learning objectives

D. time estimations

12

Which of the following is a solution of the given equation?

$4(m + 4)^2 - 4m^2 + 20 = 276$

A. 3

B. 6

C. 12

D. 24

13

If $\triangle ABD \sim \triangle DEF$ and the similarity ratio is 3:4, what is the measure of DE if $AB = 12$?

A. 9

B. 16

C. 96

D. 12

14

A cube is inscribed in a sphere such that each vertex on the cube touches the sphere. If the volume of the sphere is 972π cm³, what is the approximate volume of the cube in cubic centimeters?

A. 9

B. 10.4

C. 1125

D. 1729

15

Which of the following statements is true regarding the use of learning objectives?

A. Learning objectives are broad and general statements that describe high-level skills.

B. Learning objectives are the individual pieces that provide the basis of each lesson.

C. Learning objectives can hinder a teacher's ability to provide meaningful feedback to students.

D. Learning objectives are not necessary in order for a student to show mastery of a learning outcome.

16

A data set contains information on the hours worked in a government department. Which of the following statistics would have a unit that is NOT hours?

A. variance

B. standard deviation

C. range

D. interquartile range

17

What is the x-intercept of the given equation?

$10x + 10y = 10$

A. (1, 0)

B. (0, 1)

C. (0, 0)

D. (1, 1)

18

Which of the following is NOT included in a unit plan?

A.   learning outcomes

B.   learning objectives

C.   lesson plans

D.   curriculum map

19

A high school cross country team sent 25 percent of its runners to a regional competition. Of these, 10 percent won medals. If 2 runners earned medals, how many members does the cross country team have?

A.   8

B.   80

C.   125

D.   1250

20

Which inequality is represented by the following graph?

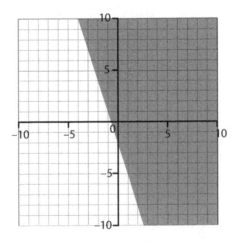

A. $y \geq -3x - 2$

B. $y \geq 3x - 2$

C. $y > -3x - 2$

D. $y \leq -3x - 2$

21

Which of the angles in the figure below are congruent?

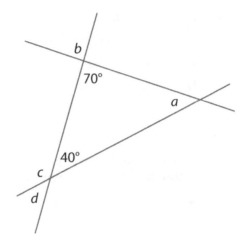

A. a and d

B. b and d

C. a and b

D. c and b

22

Fifteen DVDs are to be arranged on a shelf. 4 of the DVDs are horror films, 6 are comedies, and 5 are science fiction. In how many ways can the DVDs be arranged if DVDs of the same genre must be placed together?

A.   2,073,600

B.   6,220,800

C.   12,441,600

D.   131,216,200

23

What is the maximum value of the function $f(x) = 3\sin(x - 2) + 1$?

A.   1

B.   3

C.   4

D.   7

24

Which of the following answer choices shows the most appropriate order for the following lessons included in the unit Graphing & Analysis of Linear Equations?

1. Determine the slope of a linear equation.

2. Determine whether two lines are parallel, perpendicular, or neither by analyzing the graph of each equation.

3. Determine the x-intercept and y-intercept of a linear equation.

4. Graph an equation in slope-intercept form.

A. 1, 4, 3, 2

B. 1, 3, 4, 2

C. 2, 3, 1, 4

D. 3, 4, 2, 1

25

Points B and C are on a circle. Point A is located such that the line segments $\overline{AB}$ and $\overline{AC}$ are congruent. Which of the following could be true?

I. A is the center of the circle.

II. A is on arc $\overparen{BC}$.

III. A is outside of the circle.

A. I

B. I and II

C. I and III

D. I, II, and III

26

In a theater, there are 4,500 lower-level seats and 2,000 upper-level seats. What is the ratio of lower-level seats to total seats?

A. $\frac{4}{9}$

B. $\frac{4}{13}$

C. $\frac{9}{13}$

D. $\frac{9}{4}$

27

Cone A is similar to cone B with a scale factor of 3:4. If the volume of cone A is 54π, what is the volume of cone B?

A. 72π

B. 128π

C. 162π

D. 216π

28

A pair of 6-sided dice is rolled 10 times. What is the probability that in exactly 3 of those rolls, the sum of the dice will be 5?

A. 0.14%

B. 7.2%

C. 11.1%

D. 60%

29

If a person reads 40 pages in 45 minutes, approximately how many minutes will it take her to read 265 pages?

A. 202

B. 236

C. 265

D. 298

30

Which of the following could be the perimeter of a triangle with two sides that measure 13 and 5?

A.   24.5

B.   26.5

C.   36

D.   37

31

Which of the following tools would best aid students in a lesson on investigating the properties of the circumference, diameter, and radius of a circle?

A.   non-graphing calculator

B.   protractor

C.   base ten blocks

D.   online algebra software

32

What are the real zero(s) of the following polynomial?

$2n^2 + 2n - 12 = 0$

A.   {2}

B.   {−3, 2}

C.   {2, 4}

D.   There are no real zeros of n.

33

Which of the following is a solution to the given function?

$\cos x - \sin x = -1$

A.   0

B. $\frac{\pi}{4}$

C. $\frac{\pi}{3}$

D. $\frac{\pi}{2}$

34

What is the total number of 6-digit numbers in which each individual digit is less than 3 or greater than 6?

A.   38,880

B.   46,656

C.   80,452

D.   101,370

35

A worker was paid $15,036 for 7 months of work. If he received the same amount each month, how much was he paid for the first 2 months?

A. $2,148

B. $4,296

C. $6,444

D. $8,592

36

If the length of a rectangle is increased by 40% and its width is decreased by 40%, what is the effect on the rectangle's area?

A.   The area is the same.

B.   It increases by 16%.

C.   It increases by 20%.

D.   It decreases by 16%.

37

Which of the following assessment examples would NOT be considered a formative assessment?

A.   exit ticket

B.   class discussion

C.   unit test

D.   three-question quiz

38

In the diagram, $\overline{BD}$ is a diameter of circle O. If $\widehat{AB} = 55°$ and $\widehat{BC} = 85°$, what is the measure of angle DMC?

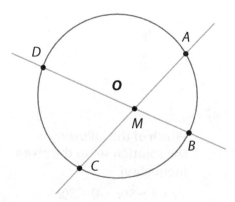

A. 70°

B. 75°

C. 80°

D. 82.5°

39

What is the value of z in the following system?

$z - 2x = 14$

$2z - 6x = 18$

A. −7

B. 3

C. 5

D. 24

40

In a class of 20 students, how many conversations must be had so that every student talks to every other student in the class?

A. 190

B. 380

C. 760

D. 6840

41

50 shares of a financial stock and 10 shares of an auto stock are valued at $1,300. If 10 shares of the financial stock and 10 shares of the auto stock are valued at $500, what is the value of 50 shares of the auto stock?

A. $30

B. $20

C. $1,300

D. $1,500

42

If lines y and z are perpendicular, which of the following is NOT a possible difference between the slopes of the two lines?

A. $-\frac{10}{3}$

B. $-\frac{1}{2}$

C. 2

D. 10

43

Which of the following statements most accurately describes backward unit design?

A. developing each lesson prior to writing the summative unit assessment

B. writing the summative unit assessment after week three of a six week unit

C. developing the summative unit assessment based on the learning outcomes prior to writing lesson plans

D. developing formative assessments prior to writing the summative unit assessment

44

The regular octagon below is inscribed within a circle. What is the approximate difference between arc $\overset{\frown}{HB}$ and arc $\overset{\frown}{CF}$?

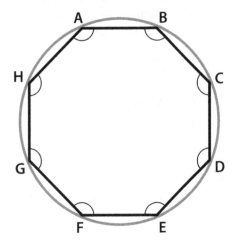

A. 45°

B. 137°

C. 320°

D. 540°

45

Which of the following statements most accurately describes the goal of a summative unit assessment?

A. measure student mastery of each learning outcome for the unit

B. measure student mastery of each learning objective covered throughout the unit

C. measure student mastery of all of learning outcomes for the course

D. measure student mastery of learning outcomes selected by the teacher

46

Simplify: $\dfrac{(3x^2y^2)^2}{3^3x^{-2}y^3}$

A. $3x^6y$

B. $\dfrac{x^6y}{3}$

C. $\dfrac{x^4}{3y}$

D. $\dfrac{3x^4}{y}$

47

Which of the following is the solution set to the given inequality?

$2x + 4 \geq 5(x - 4) - 3(x - 4)$

A. $(-\infty, \infty)$

B. $(-\infty, 6.5]$

C. $[6.5, -\infty)$

D. $(-\infty, 6.5) \cup (6.5, \infty)$

48

What is the axis of symmetry for the given parabola?

$y = -2(x + 3)^2 + 5$

A. $y = 3$

B. $x = -3$

C. $y = -3$

D. $x = 3$

49

Which statement about the following set is true?

$\{60, 5, 18, 20, 37, 37, 11, 90, 72\}$

A. The median and the mean are equal.

B. The mean is less than the mode.

C. The mode is greater than the median.

D. The median is less than the mean.

50

Which of the following statements must be true for triangle *ABC*?

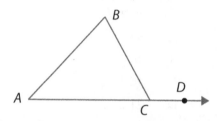

I.	$AC + CB > AB$
II.	$m\angle BCD = m\angle A + m\angle B$
III.	$m\angle BCD + m\angle BCA = 90°$
IV.	$m\angle A = m\angle C$

A.   I only

B.   I and II

C.   II and IV

D.   I, II, and III

51

What is the greatest number of complex roots a 17th degree polynomial can have?

A.   8

B.   17

C.   16

D.   16*i*

52

Simplify: $\dfrac{3 + \sqrt{3}}{4 - \sqrt{3}}$

A. $\dfrac{13}{15}$

B. $\dfrac{15 + 7\sqrt{3}}{13}$

C. $\dfrac{15}{19}$

D. $\dfrac{15 + 7\sqrt{3}}{19}$

53

A company interviewed 21 applicants for a recent opening. Of these applicants, 7 wore blue and 6 wore white, while 5 applicants wore both blue and white. What is the number of applicants who wore neither blue nor white?

A.   1

B.   6

C.   12

D.   13

54

Which of the following statements describes the difference between summative and formative assessments?

A.   Summative assessments are often high stakes and high point value; formative assessments are often low stakes and low point value.

B.   Summative assessments are often low stakes and low point value; formative assessments are often high stakes and high point value.

C.   Summative assessments aid the teacher in determining student misunderstandings while formative assessments do not.

D.   Summative assessments are given for the sole purpose of a grade while formative assessments help the teacher give meaningful and specific feedback.

55

W, X, Y, and *Z* lie on a circle with center *A*. If the diameter of the circle is 75, what is the sum of $\overline{AW}$, $\overline{AX}$, $\overline{AY}$, and $\overline{AZ}$?

A. 75

B. 300

C. 150

D. 106.5

56

A wedge from a cylindrical piece of cheese was cut as shown. If the entire wheel of cheese weighed 73 pounds before the wedge was removed, what is the approximate remaining weight of the cheese?

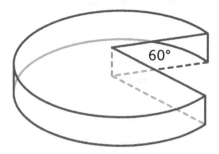

A. 12.17 pounds

B. 37.00 pounds

C. 60.83 pounds

D. 66.92 pounds

57

asymptote of the function

$f(x) = \frac{x+4}{-2x-6}$?

A. $y = \frac{1}{2}$

B. $y = -2$

C. $x = 3$

D. $x = -3$

58

The mean of 13 numbers is 30. The mean of 8 of these numbers is 42. What is the mean of the other 5 numbers?

A. 5.5

B. 10.8

C. 16.4

D. 21.2

59

What are the roots of the equation $y = 16x^3 - 48x^2$?

A. $\left\{ \frac{3+i\sqrt{5}}{2}, \frac{3-i\sqrt{5}}{2} \right\}$

B. $\{0, 3, -3\}$

C. $\{0, 3i, -3i\}$

D. $\{0, 3\}$

60

Which of the following strategies is most appropriate when it is necessary for a teacher to give direct instruction?

A. a full class period of lecture

B. I do, we do, you do

C. assign the material for homework to avoid using class time to give direct instruction

D. think, pair, share

61

$\overline{MN}$ is the diameter of circle *O*. If the coordinates of *M* are (4, 5) and the coordiantes of *N* are (−12, −11), what is the equation for circle *O*?

A. $(x + 4)^2 + (y + 3)^2 = 100$

B. $(x - 4)^2 + (y - 3)^2 = 10$

C. $(x + 4)^2 + (y - 3)^2 = 10$

D. $(x - 4)^2 + (y + 3)^2 = 100$

62

If angles *a* and *b* are congruent, what is the measurement of angle *c*?

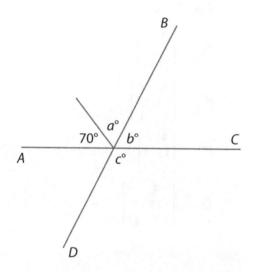

A. 70°

B. 125°

C. 110°

D. 55°

63

If the volume of a cube is 343 cubic meters, what is the cube's surface area?

A. 49 m²

B. 84 m²

C. 196 m²

D. 294 m²

64

Find $f(x) = (j \circ t)(x)$ if $j(x) = \frac{1}{3}x - 2$ and $t(x) = \frac{1}{2}x - 3$.

A. $f(x) = \frac{1}{6}x^2 + 6$

B. $f(x) = \frac{1}{6}x^2 - 5x + 6$

C. $f(x) = \frac{1}{6}x - 4$

D. $f(x) = \frac{1}{6}x - 3$

65

Which of the following assessment examples would be considered a formal assessment?

A. exit ticket

B. unit test written by the teacher

C. standardized unit test

D. free response question written by the teacher

66

In the figure below, there are six line segments that terminate at point *O*. If segment $\overline{DO}$ bisects $\angle AOF$ and segment $\overline{BO}$ bisects $\angle AOD$, what is the value of $\angle AOF$?

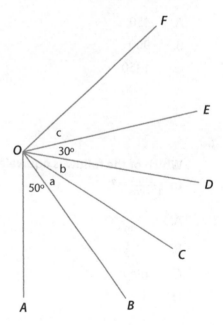

A. 140°

B. 160°

C. 170°

D. 200°

67

Given that $\overline{ED}$ and $\overline{AC}$ are parallel, $\triangle ABC$ is a 45–45–90 triangle, and D is the midpoint of $\overline{AB}$, what is the area of the unshaded parts of the triangle below?

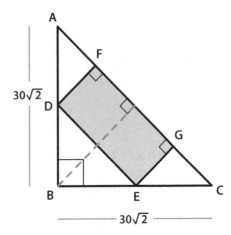

A.   450

B.   900

C.   1350

D.   1800

68

Which of the follow is equivalent to $\frac{\sin x}{1 - \cos x}$?

A. $\frac{1 + \cos x}{\sin x}$

B. $\frac{\sin x}{\cos x}$

C. $\tan x$

D.   1

69

In the *xy*-coordinate plane, how many points have a distance of four from the origin?

A.   0

B.   2

C.   4

D.   ∞

70

If $B = \begin{bmatrix} 6 & 4 \\ 8 & -2 \\ 5 & -3 \end{bmatrix}$ and $C = \begin{bmatrix} -2 & 5 \\ 7 & 1 \\ -4 & 4 \end{bmatrix}$, find $B + C$.

A. $\begin{bmatrix} 4 & 9 \\ 15 & -1 \\ -1 & -1 \end{bmatrix}$

B. $\begin{bmatrix} 4 & 9 \\ 15 & -1 \\ 1 & 1 \end{bmatrix}$

C. $\begin{bmatrix} 4 & 9 \\ 15 & 1 \\ 1 & 1 \end{bmatrix}$

D. $\begin{bmatrix} -4 & -9 \\ -15 & 1 \\ -1 & -1 \end{bmatrix}$

71

If the smallest angle in a non-right triangle is 20° and the shortest side is 14, what is the length of the largest side if the largest angle is 100°?

A.   12.78

B.   34.31

C.   40.31

D.   127.81

72

A bag contains 6 blue, 8 silver, and 4 green marbles. Two marbles are drawn from the bag. What is the probability that the second marble drawn will be green if replacement is not allowed?

A. $\frac{2}{9}$

B. $\frac{4}{17}$

C. $\frac{11}{17}$

D. $\frac{7}{9}$

73

What is the approximate diagonal length of square QTSR shown in the figure below?

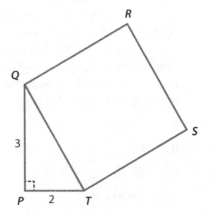

A.   3

B.   3.6

C.   5.1

D.   13

74

Rectangular water tank A is 5 feet long, 10 feet wide, and 4 feet tall. Rectangular tank B is 5 feet long, 5 feet wide, and 4 feet tall. If the same amount of water is poured into both tanks and the height of the water in Tank A is 1 foot, how high will the water be in Tank B?

A.   1 foot

B.   2 feet

C.   3 feet

D.   4 feet

75

What is the domain of the inequality $\left|\frac{x}{8}\right| \geq 1$?

A. $(-\infty, \infty)$

B. $[8, \infty)$

C. $(-\infty, -8]$

D. $(-\infty, -8] \cup [8, \infty)$

76

The line of best fit is calculated for a data set that tracks the number of miles that passenger cars traveled annually in the US from 1960 to 2010. In the model, $x = 0$ represents the year 1960, and y is the number of miles traveled in billions. If the line of best fit is $y = 0.0293x + 0.563$, approximately how many additional miles were traveled for every 5 years that passed?

A.   0.0293 billion

B.   0.1465 billion

C.   0.563 billion

D.   0.710 billion

77

Two spheres are tangent to each other. One has a volume of 36π, and the other has a volume of 288π. What is the greatest distance between a point on one of the spheres and a point on the other sphere?

A.   6

B.   9

C.   18

D.   63

78

What are the vertices of an ellipse defined by the equation:
$$\frac{(x-8)^2}{100} + \frac{(y-4)^2}{25} = 1?$$

A. $(18, 4)$ and $(-2, 4)$

B. $(8, 9)$ and $(8, -1)$

C. $(8, 14)$ and $(8, -6)$

D. $(13, 4)$ and $(3, 4)$

79

Line *a* is tangent to circle *X* below. If $\overline{XY}$ and $\overline{WY}$ are congruent, what is the measurement of ∠*WZX*?

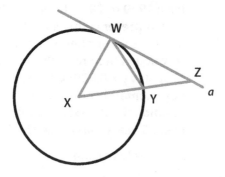

A.   30°

B.   45°

C.   60°

D.   90°

80

If the perimeter of an equilateral triangle is 30 inches, what is the altitude of the triangle?

A.   5

B. $5\sqrt{3}$

C.   10

D. $15\sqrt{3}$

81

If the water in a reservoir decreases every day by 4%, by how much will the water decrease over a 7-day week?

A.   24.9%

B.   28.0%

C.   96.0%

D.   131.6%

82

A spinner has 10 equal-sized sectors numbered 1 to 10. If the spinner lands on the number chosen by the player, he receives $10. If the player chooses an even number, and an even number is spun, he wins $3. If a bet of $2 is made that the spinner will land on the number 8, what is the expected value of this bet?

A.   −$0.50

B. $0

C. $0.20

D. $0.50

83

If $\mathbf{D} = \begin{bmatrix} 3 & 7 \\ 4 & 9 \end{bmatrix}$, find det(**D**).

A.   −2

B.   −1

C.   1

D.   2

84

Find the approximate value of *x* in the triangle below.

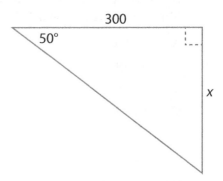

A.   229

B.   300

C.   357

D.   400

85

Which of the following is an alternative to a summative unit assessment for a unit in which students participated in a community garden project in order to learn about the concepts of area and perimeter?

A. create a video summarizing the project and the ultimate learning outcomes

B. create a portfolio showing each stage of the process and the ultimate learning outcomes

C. give an oral presentation summarizing the project and the ultimate learning outcomes

D. all of the above

86

What is the approximate value of a in the triangle below if b is $\frac{7}{8}$ the value of a?

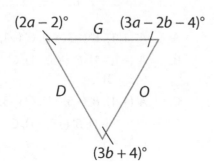

A. ≈ 22

B. ≈ 30

C. ≈ 31

D. ≈ 42

87

Find the 12th term of the following sequence.

$-57, -40, -23, -6...$

A. 57

B. 79

C. 113

D. 130

88

What transformation is created by the -3 in the graph of $y = -3|x - 2| + 2$?

A. The -3 moves the vertex down 3 and reflects the graph over the x-axis.

B. The -3 moves the vertex to the left 3 and widens the graph.

C. The -3 makes the graph wider and reflects it over the x-axis.

D. The -3 makes the graph narrower and reflects the graph over the x-axis.

89

Which of the following is an example of a well-structured guiding question for a teacher to ask during a lesson covering the concepts of area and perimeter?

A. What is the most important thing you know about area and perimeter?

B. Can you draw a model/picture to illustrate the definitions of area and perimeter?

C. What is the area of a square?

D. What is the perimeter of a square?

90

If $y = 2x^2 + 12x - 3$ is written if the form $y = a(x - h)^2 + k$, what is the value of k?

A.   −3

B.   −15

C.   −18

D.   −21

91

If $\mathbf{D} = \begin{bmatrix} 3 & 7 \\ 4 & 9 \end{bmatrix}$, find: $\mathbf{D}^{-1}$

A. $\begin{bmatrix} 3 & 7 \\ 4 & 9 \end{bmatrix}$

B. $\begin{bmatrix} 9 & 7 \\ 4 & 3 \end{bmatrix}$

C. $\begin{bmatrix} -9 & 7 \\ 4 & -3 \end{bmatrix}$

D. $\begin{bmatrix} -9 & -7 \\ -4 & -3 \end{bmatrix}$

92

Which of the following is an example of a well-structured guiding question for a teacher to ask a student who is struggling with the first step in graphing a quadratic equation?

A.   What is the vertex?

B.   What are the intercepts of the graph?

C.   How might you construct an x-y table to determine coordinate pairs?

D.   Is the parabola concave up or concave down?

93

What are the zeros of $\left(\dfrac{g}{h}\right)(k)$ if $g(k) = -3k^2 - k$ and $h(k) = -2k - 4$?

A. $\left\{0, \dfrac{1}{3}\right\}$

B. $\{-2\}$

C. $\{0\}$

D. $\left\{-2, 0, \dfrac{1}{3}\right\}$

94

If $\triangle ABC$ is rotated counterclockwise 180° about point A, what are the coordinates of the new triangle?

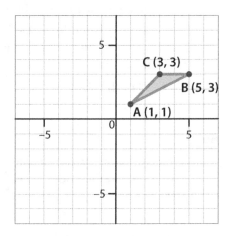

A. $A'\,(1, 1)$, $B'\,(-3, -1)$, $C'\,(-1, -1)$

B. $A'\,(-1, -1)$, $B'\,(-5, -1)$, C' $(-3, -3)$

C. $A'\,(1, 1)$, $B'\,(-5, -1)$, $C'\,(-3, -3)$

D. $A'\,(-1, -1)$, $B'\,(-3, -1)$, C' $(-1, -1)$

95

If the surface area of a cylinder with radius of 4 feet is 48π square feet, what is its volume?

A. 1π ft.³

B. 16π ft.³

C. 32π ft.³

D. 64π ft.³

96

If an employee who makes $37,500 per year receives a 5.5% raise, what is the employee's new salary?

A. $35,437.50

B. $35,625

C. $39,375

D. $39,562.50

97

New York had the fewest months with less than 3 inches of rain in every year except:

Number of Months with 3 or Fewer Than 3 Inches of Rain

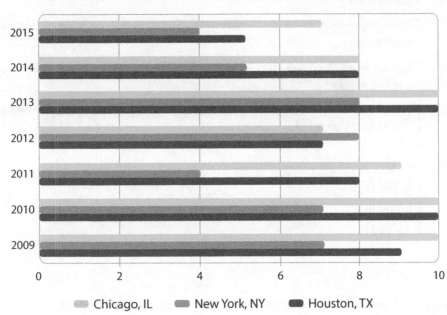

A.   2012

B.   2013

C.   2014

D.   2015

98

600 people between the ages of 15 and 45 were polled regarding their use of social media. 200 people from each age group were part of the study. The results are listed in the relative frequency table below.

Hours Per Day Spent on Social Media				
Age (years)	Less than 2 hours	2 to 4 hours	More than 4 Hours	Total
15 – 25	0.15	0.40	0.45	1.0
25 – 35	0.52	0.28	0.20	1.0
35 – 45	0.85	0.10	0.05	1.0
Total	0.51	0.26	0.23	1.0

Which of the statements are true?

I.   Of people 25 to 35 years old, 20% spend more than 4 hours per day on social media.

II. Of the population of 15 to 45-year-olds, 26% spend 2 – 4 hours a day on social media.

III. Of people between the ages of 15 and 35, 67% spend 0 – 2 hours per day on social media.

IV. 5 people who reported using social media more than 4 hours per week were 35 to 45 years old.

A.   I and II

B.   I, II, and IV

C.   I, II, and III

D.   II, III, and IV

99

The population of a city in 2008 was 1.25 million people. If the population is decreasing by 5% annually, in what year will the population reach 1 million?

A.   2011

B.   2012

C.   2013

D.   2014

100

What is the value of z in the following equation?

$\log_7(-2z) = 0$

A. $-\frac{1}{2}$

B.   0

C. $\frac{1}{2}$

D. $\frac{7}{2}$

Answer Key

1)

C. FOIL and combine like terms.

$(5 + 2i)(3 + 4i)$

$= 15 + 6i + 20i + 8i^2$

$= 15 + 6i + 20i + (8)(-1)$

$= 15 + 6i + 20i - 8$

$= \textbf{7 + 26}\textbf{\textit{i}}$

2)

D. Rewrite the bases so they are the same, then set the exponents equal and solve.

$16^{x+10} = 8^{3x}$

$(2^4)^{x+10} = (2^3)^{3x}$

$2^{4x+40} = 2^{9x}$

$4x + 40 = 9x$

$\textbf{\textit{x} = 8}$

3)

B. Use the midpoint formula to find point B.

$M_x: \frac{(7+x)}{2} = -3$

$x = -13$

$M_y: \frac{(12+y)}{2} = 10$

$y = 8$

$B = \textbf{(-13, 8)}$

4)

A. Find the amount the state will spend on infrastructure and education, and then find the difference.

infrastructure $= 0.2(3{,}000{,}000{,}000) =$ 600,000,000

education $= 0.18(3{,}000{,}000{,}000) =$ 540,000,000

$600{,}000{,}000 - 540{,}000{,}000 =$ **$60,000,000**

5)

A. Use the formula for inversely proportional relationships to find k and then solve for s.

$sn = k$

$(65)(250) = k$

$k = 16{,}250$

$s(325) = 16{,}250$

$s = \textbf{50}$

6)

D. Solve each equation for y and find the equation with a power of 1.

$\sqrt[3]{y} = x \rightarrow y = x^3$

$\sqrt[3]{x} = y \rightarrow y = \sqrt[3]{x}$

$\sqrt[3]{y} = x^2 \rightarrow y = x^6$

$y = \sqrt[3]{x^3} \rightarrow \textbf{\textit{y} = \textit{x}}$

7)

B. On a normal distribution curve, the group of students with a height between 5.25 and 5.5 feet is one standard deviation from the mean. More students will be in this group than any of the others, which are all more than one standard deviation from the mean.

8)

C. Find the slope of line b, take the negative reciprocal to find the slope of a, and test each point.

$(x_1, y_1) = (-100, 100)$

$(x_2, y_2) = (-95, 115)$

$m_b = \frac{115 - 100}{-95 - (-100)} = \frac{15}{5} = 3$

$m_a = -\frac{1}{3}$

$(104, 168): \frac{100 - 168}{-100 - (104)} = \frac{1}{3}$

$(-95, 115): \frac{100 - 115}{-100 - (-95)} = 3$

$(-112, 104): \frac{100 - 104}{-100 - (-112)} = -\frac{1}{3}$

$(-112, -104): \frac{100 - (-104)}{-100 - (-112)} = 17$

9)

A. Incorrect. Learning outcomes describe high-level skills that include the ability to analyze, evaluate, and create based on the content learned during the unit or course.

B. Incorrect. Learning outcomes describe high-level skills that include the ability to analyze, evaluate, and create based on the content learned during the unit or course.

C. Incorrect. Learning outcomes describe high-level skills that include the ability to analyze, evaluate, and create based on the content learned during the unit or course.

D. **Correct.** Memorization is a low-level skill not typically described by a learning outcome.

10)

A. If the data is skewed right, the set includes extremes values that are to the right, or high. The median is unaffected by these high values, but the mean includes these high values and would therefore be greater.

11)

A. Incorrect. A curriculum map breaks the year into units.

B. Incorrect. A curriculum map shows the learning outcomes associated with each unit.

C. **Correct.** A curriculum map does not specifically show every learning objective to be covered during the year.

D. Incorrect. A curriculum map gives time estimations for each unit in order to ensure that all necessary content is covered.

12)

B. Plug each value into the equation.

$4(3 + 4)^2 - 4(3)^2 + 20 = 180 \neq 276$

$4(6 + 4)^2 - 4(6)^2 + 20 = \mathbf{276}$

$4(12 + 4)^2 - 4(12)^2 + 20 = 468 \neq 276$

$4(24 + 4)^2 - 4(24)^2 + 20 = 852 \neq 276$

13)

B. Set up a proportion and solve.

$\frac{AB}{DE} = \frac{3}{4}$

$\frac{12}{DE} = \frac{3}{4}$

$3(DE) = 48$

$\mathbf{DE = 16}$

14)

C. Use the formula for the volume of a sphere to find its radius.

$V = \frac{4}{3}\pi r^3$

$972\pi = \frac{4}{3}\pi r^3$

$r = 9$

Use the super Pythagorean theorem to find the side of the cube.

$$d^2 = a^2 + b^2 + c^2$$

$$18^2 = 3s^2$$

$$s \approx 10.4$$

Use the length of the side to find the volume of the cube.

$$V = s^3$$

$$V \approx (10.4)^3$$

$$V \approx \mathbf{1{,}125}$$

15)

A. Incorrect. Learning outcomes are broad and general statements that describe high-level skills.

B. **Correct.** Objectives are the building blocks of each lesson.

C. Incorrect. Objectives are an aid in providing meaningful and specific feedback to students.

D. Incorrect. Typically, learning objectives allow a teacher to scaffold each lesson in such a way that ultimately leads students to mastery of a learning outcome.

16)

A. Variance would have a unit of hours squared, not hours.

17)

A. Plug 0 in for y and solve for x.

$$10x + 10y = 10$$

$$10x + 10(0) = 10$$

$$x = 1$$

The x-intercept is at **(1, 0)**.

18)

A. Incorrect. A unit plan includes learning outcomes, learning objectives, pre-assessment, summative unit assessment, and lesson plans.

B. Incorrect. A unit plan includes learning outcomes, learning objectives, pre-assessment, summative unit assessment, and lesson plans.

C. Incorrect. A unit plan includes learning outcomes, learning objectives, pre-assessment, summative unit assessment, and lesson plans.

D. **Correct.** The curriculum map is used as a pacing guide and gives unit titles and time estimations; however, the curriculum map is not necessarily included in the unit plan.

19)

B. Work backwards to find the number of runners in the competition (c) and then the number of runners on the team (r).

$$\frac{2}{c} = \frac{10}{100}$$

$$c = 20$$

$$\frac{20}{r} = \frac{25}{100}$$

$$r = \mathbf{80}$$

20)

Eliminate answer choices that don't match the graph.

A. Correct.

B. The graph has a negative slope while this inequality has a positive slope.

C. The line on the graph is solid, so the inequality should include the "or equal to" symbol.

D. The shading is above the line, meaning the inequality should be "y is greater than."

21)

C. Find the measure of each angle.

$$m\angle a = 180 - (70 + 40) = 70°$$

$$m\angle b = 70°$$

$$m\angle c = 180 - 40 = 140°$$

$$m\angle d = 40°$$

$\angle a \cong \angle b$

22)

C. Use the fundamental counting principle to determine how many ways the DVDs can be arranged within each category and how many ways the 3 categories can be arranged.

ways to arrange horror = 4! = 24

ways to arrange comedies = 6! = 720

ways to arrange science fiction = 5! = 120

ways to arrange categories = 3! = 6

(24)(720)(120)(6) = **12,441,600**

23)

C. This sine function has an amplitude of 3 and has been shifted up 1, so it has a maximum value of **4**.

24)

A. Incorrect. Students should understand how to solve for the slope and intercepts before graphing the equation.

B. **Correct.** Students should understand how to solve for the slope and intercepts and how to graph the equation prior to the more abstract idea of parallel and perpendicular lines.

C. Incorrect. Understanding the more abstract idea of parallel and perpendicular lines should be the last lesson in this sequence rather than the first.

D. Incorrect. Students should understand how to solve for and analyze the slope prior first rather than last.

25)

D. As shown in the figure, A can be placed inside, on, or outside the circle.

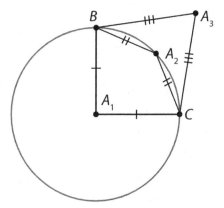

26)

C.

total seats = 4,500 + 2,000

$\dfrac{\text{lower seats}}{\text{all seats}} = \dfrac{4,500}{6,500} = \dfrac{9}{13}$

27)

B. Set up a proportion. Cube the scale factor when calculating volume.

$\dfrac{54\pi}{x} = \dfrac{3^3}{4^3}$

$x = 128\pi$

28)

B. Use the equation for Bernoulli trials (binomial distribution).

$P = {}_nC_r(p^r)(q^{n-r})$

$n = 10$

$r = 3$

$p = \dfrac{4}{36} = \dfrac{1}{9}$

$q = \dfrac{8}{9}$

$P = {}_{10}C_3\left(\dfrac{1}{9}\right)^3\left(\dfrac{8}{9}\right)^7 = 0.072 = \textbf{7.2\%}$

29)

D. Write a proportion and then solve for *x*.

$\dfrac{40}{45} = \dfrac{265}{x}$

$40x = 11,925$

$x = 298.125 \approx \mathbf{298}$

30)

B.   Use the triangle inequality theorem to find the possible values for the third side, then calculate the possible perimeters.

$13 - 5 < s < 13 + 5$

$8 < s < 18$

$13 + 5 + 8 < P < 13 + 5 + 18$

$26 < P < 36$

26.5 is the only answer choice in this range.

31)

A.   Incorrect. While students could use a non-graphing calculator to aid in the arithmetic, this is not the best tool listed in the answer choices.

B.   Correct. Students can use a protractor to construct models and illustrations of various circles to aid in the investigation of the properties of circumference, diameter, and radius.

C.   Incorrect. Base ten blocks are not necessary tools for this concept.

D.   Incorrect. Online algebra software is an unnecessary tool for this concept.

32)

B.   Factor the trinomial and set each factor equal to 0.

$2n^2 + 2n - 12 = 0$

$2(n^2 + n - 6) = 0$

$2(n + 3)(n - 2) = 0$

$\mathbf{n = -3}$ and $\mathbf{n = 2}$

33)

D.   Plug each value into the function and evaluate.

$\cos 0 - \sin 0 = 1 \neq -1$

$\cos \frac{\pi}{4} - \sin \frac{\pi}{4} = \frac{1}{2} - \frac{\sqrt{3}}{2} \neq -1$

$\cos \frac{\pi}{3} - \sin \frac{\pi}{3} = \frac{(1 - \sqrt{3})}{2} \neq -1$

$\cos \frac{\pi}{2} - \sin \frac{\pi}{2} = \mathbf{-1}$

34)

A.   There are six digits that can be used to make up the 6-digit number: 0, 1, 2, 7, 8, and 9. However, 0 cannot be used for the first digit. Use the fundamental counting principle: (5)(6)(6)(6)(6)(6) = **38,880**.

35)

B.   Write a proportion and then solve for x.

$\frac{15,036}{7} = \frac{x}{2}$

$7x = 30,072$

$x = \mathbf{4,296}$

36)

D.   Use the formula for the area of a rectangle to find the increase in its size.

$A = lw$

$A = (1.4l)(0.6w)$

$A = 0.84lw$

The new area will be 84% of the original area, a decrease of **16%**.

37)

A.   Incorrect. Exit tickets are a form of formative assessment.

B.   Incorrect. Class discussions can be considered a form of formative assessment.

C.   Correct. Unit tests are a form of summative assessment.

D.   Incorrect. Brief quizzes are a form of formative assessment.

38)

B.   Find m∠DMC by averaging the lengths of $\overset{\frown}{DC}$ and $\overset{\frown}{AB}$.

$\overset{\frown}{AB} = 55°$

$\overset{\frown}{DC} = 180° - 85° = 95°$

$m\angle DMC = \frac{55 + 95}{2} = \mathbf{75°}$

39)

D. Solve the system using substitution.

$z - 2x = 14 \rightarrow z = 2x + 14$

$2z - 6x = 18$

$2(2x + 14) - 6x = 18$

$4x + 28 - 6x = 18$

$-2x = -10$

$x = 5$

$z - 2(5) = 14$

$\mathbf{z = 24}$

40)

A. Use the combination formula to find the number of ways to choose 2 people out of a group of 20.

$C(20, 2) = \frac{20!}{2!\,18!} = \mathbf{190}$

41)

D. Set up a system of equations and solve using elimination.

f = the cost of a financial stock

a = the cost of an auto stock

$50f + 10a = 1300$

$10f + 10a = 500$

$50f + 10a = 1300$

$\underline{+\; -50f - 50a = -2500}$

$-40a = -1,200$

$a = 30$

$50(30) = \mathbf{1,500}$

42)

B. Two perpendicular lines have slopes m and $-\frac{1}{m}$. Their difference is $m - (-\frac{1}{m}) = m + \frac{1}{m}$. To determine if an answer choice is possible, set this quantity equal to each choice and solve.

Note that answer choice B does not work. B. $-\frac{1}{2}$

$m + \frac{1}{m} = \frac{-1}{2}$

$2m^2 + 2 = -1m$

$2m^2 + 1m + 2 = 0$

$m = \frac{-1 \pm \sqrt{1 - 4(2)(2)}}{4}$

There is no real solution, so $-\frac{1}{2}$ cannot be a difference between the two slopes.

43)

A. Incorrect. The summative unit assessment should be designed before writing lessons.

B. Incorrect. The summative unit assessment should be designed prior to implementation of any lessons within the unit.

C. **Correct.** The summative unit assessment should be designed based on the unit's learning outcomes prior to writing or implementing and of the lessons.

D. Incorrect. The summative unit assessment should be designed prior to writing lessons which include formative assessments.

44)

A. Find the measure of each arc and then find the difference in the two segments.

$each\ arc = \frac{360°}{8} = 45°$

The difference between CF and HB is one arc, or **45°**.

45)

A. **Correct.** The goal of the summative unit assessment is to measure mastery of each learning outcome included in the unit of study.

B. Incorrect. Formative assessments throughout the unit are intended to check for mastery of specific learning objectives.

C. Incorrect. An end-of-course test should measure mastery of all of the outcomes for a particular course

D. Incorrect. The summative unit assessment should measure mastery of all necessary unit outcomes, not just the outcomes selected by the teacher.

46)

B. Use the rules of exponents to simplify the expression.

$$\frac{(3x^2y^2)^2}{3^3x^{-2}y^3} = \frac{3^2x^4y^4}{3^3x^{-2}y^3} = \frac{x^6y}{3}$$

47)

A. Simplify the inequality.

$2x + 4 \geq 5x - 20 - 3x + 12$

$2x + 4 \geq 2x - 8$

$4 \geq -8$

Since the inequality is always true, the solution is all real numbers, $(-\infty, \infty)$.

48)

B. The axis of symmetry will be a vertical line that runs through the vertex, which is the point $(-3, 5)$. The line of symmetry is $x = -3$.

49)

D. {5, 11, 18, 20, **37**, 37, 60, 72, 90}

median = 37

mode = 37

mean

$= \frac{60 + 5 + 18 + 20 + 37 + 37 + 11 + 90 + 72}{9}$

$= 38.89$

The median is less than the mean.

50)

B. I. True: The sum of any two sides of a triangle must always be greater than the third side.

II. True: The exterior angle of a triangle is always equal to the sum of the opposite interior angles.

III. False: $\angle BCD$ and $\angle BCA$ are a linear pair, so they sum to 180°, not 90°.

IV. False: This cannot be determined from the information in the figure.

51)

C. Complex solutions always come in pairs. Therefore, the number of possible complex solutions is the greatest *even* number equal to or less than the power of the polynomial. A 17th degree polynomial can have at most **16** complex roots.

52)

B. Multiply by the complex conjugate and simplify.

$$\frac{3 + \sqrt{3}}{4 - \sqrt{3}} \left(\frac{4 + \sqrt{3}}{4 + \sqrt{3}} \right)$$

$$= \frac{12 + 4\sqrt{3} + 3\sqrt{3} + 3}{16 - 4\sqrt{3} + 4\sqrt{3} - 3} = \frac{15 + 7\sqrt{3}}{13}$$

53)

D. Set up an equation to find the number of people wearing neither white nor blue. Subtract the number of people wearing both colors so they are not counted twice.

$21 = 7 + 6 + neither - 5$

neither = **13**

54)

A. **Correct.** Summative assessments are often high stakes and high point value whole formative assessments are often low stakes and low point value.

B. Incorrect. Summative assessments are often high stakes and high point value; formative assessments are often low stakes and low point value.

C. Incorrect. Both summative and formative assessments aid the teacher in determining misunderstandings.

D. Incorrect. Both summative and formative assessments help the teacher give meaningful and specific feedback; no assessment should be given for the sole purpose of a grade.

$16x^2(x - 3) = 0$

$x = 0$ and $x = 3$

55)

C. All the points lie on the circle, so each line segment is a radius. The sum of the 4 lines will be 4 times the radius.

$r = \frac{75}{2} = 37.5$

$4r = \textbf{150}$

56)

C. Set up a proportion to find the weight of the removed wedge.

$\frac{60°}{x \text{ lb.}} = \frac{360°}{73 \text{ lb.}}$

$x \approx 12.17 \text{ lb.}$

Subtract the removed wedge from the whole to find the weight of the remaining piece.

$73 - 12.17 = \textbf{60.83}$

57)

D. Find where the denominator equals 0.

$-2x - 6 = 0$

$x = \textbf{-3}$

58)

B. Find the sum of the 13 numbers whose mean is 30.

$13 \times 30 = 390$

Find the sum of the 8 numbers whose mean is 42.

$8 \times 42 = 336$

Find the sum and mean of the remaining 5 numbers.

$390 - 336 = 54$

$\frac{54}{5} = \textbf{10.8}$

59)

D. Factor the equation and set each factor equal to 0.

$y = 16x^3 - 48x^2$

60)

A. Incorrect. Teachers should avoid spending full class periods in direct instruction.

B. **Correct.** I do, we do, you do is an instructional strategy to use when direct instruction is necessary.

C. Incorrect. A teacher should not assign material for homework in order to avoid spending time in class covering the material.

D. Incorrect. This is considered a formative assessment strategy.

61)

A. Use the midpoint formula to find the center of the circle and the distance formula to find its radius.

$M_x = \frac{4 + (-12)}{2} = -4$

$M_y = \frac{5 + (-11)}{2} = -3$

$M = (-4, -3)$

$r = \frac{1}{2}\sqrt{(4 - (-8))^2 + (5 - (-11))^2} = 10$

Use the center and radius to write the equation for the circle.

$(x - h)^2 + (y - k)^2 = r^2$

$\textbf{(x + 4)}^2 + \textbf{(y + 3)}^2 = \textbf{100}$

62)

B. Use the two sets of linear angles to find b and then c.

$a = b$

$a + b + 70 = 180$

$2a + 70 = 180$

$a = b = 55°$

$b + c = 180°$

$55 + c = 180$

$c = \textbf{125°}$

63)

D. Use the volume to find the length of the cube's side.

$V = s^3$

$343 = s^3$

$s = 7 \, \text{m}$

Find the area of each side and multiply by 6 to find the total surface area.

$7(7) = 49 \, \text{m}$

$49(6) = \mathbf{294 \, m^2}$

64)

D. To find the compound function, substitute $t(x)$ for x in $j(x)$.

$(j \circ t)(x) = \frac{1}{3}\left(\frac{1}{2}x - 3\right) - 2 = \mathbf{\frac{1}{6}x - 3}$

65)

A. Incorrect. Exit tickets are an informal assessment strategy.

B. Incorrect. Teacher-written unit tests are an informal assessment strategy.

C. **Correct.** Standardized tests are a form of formal assessment.

D. Incorrect. Free response questions, when not a part of a standardized test, are considered an informal assessment strategy.

66)

D. Set up a system of equations.

OD bisects $\angle AOF$: $50 + a + b = 30 + c$

BO bisects $\angle AOD$: $50 = a + b$

Substitute and solve.

$50 + 50 = 30 + c$

$c = 70$

Add each angle to find $m\angle AOF$.

$\angle AOF = 50° + a° + b° + 30° + c°$

$\angle AOF = 50° + 50° + 30° + 70°$

$\angle AOF = \mathbf{200°}$

67)

A. Find the area of the triangle using the legs as the base and height.

$A_T = \frac{(30\sqrt{2})(30\sqrt{2})}{2} = 900$

Find DE, which is the hypotenuse of $\triangle BDE$.

$DE = \sqrt{(DB)^2 + (BE)^2}$

$= \sqrt{(15\sqrt{2})^2 + (15\sqrt{2})^2}$

$= 30$

Find DF, which is the leg of $\triangle ADF$.

$DF = \frac{15\sqrt{2}}{\sqrt{2}}$

$= 15$

FInd the area of the rectangle and subtract from the area of the triangle.

$A = 900 - (30)(15)$

$= \mathbf{450}$

68)

A. Use trigonometric identities.

$\frac{\sin x}{1 - \cos x} \times \frac{1 + \cos x}{1 + \cos x}$

$\frac{(\sin x)(1 + \cos x)}{(1 - \cos^2 x)}$

$\frac{(\sin x)(1 + \cos x)}{\sin^2 x}$

$\mathbf{\frac{(1 + \cos x)}{\sin x}}$

69)

D. There are an infinite number of points with distance four from the origin, all of which lie on a circle centered at the origin with a radius of 4.

70)

B. Add the corresponding parts of each matrix.

$\begin{bmatrix} 6 & 4 \\ 8 & -2 \\ 5 & -3 \end{bmatrix} + \begin{bmatrix} -2 & 5 \\ 7 & 1 \\ -4 & 4 \end{bmatrix} = \begin{bmatrix} \mathbf{4} & \mathbf{9} \\ \mathbf{15} & \mathbf{-1} \\ \mathbf{1} & \mathbf{1} \end{bmatrix}$

71)

C. Use the law of sines.

$\frac{\sin 20°}{14} = \frac{\sin 100°}{x}$

$x = \frac{14(\sin 100°)}{\sin 20°}$

$x = 40.31$

72)

A. Find the probability that the second marble will be green if the first marble is blue, silver, or green, and then add these probabilities together.

P(first blue and second green) =
P(blue) × P(green|first blue) = $\frac{6}{18}$ × $\frac{4}{17}$ = $\frac{4}{51}$

P(first silver and second green) =
P(silver) × P(green|first silver) = $\frac{8}{18}$ × $\frac{4}{17}$ = $\frac{16}{153}$

P(first green and second green) =
P(green) × P(green|first green) = $\frac{4}{18}$ × $\frac{3}{17}$ = $\frac{2}{51}$

P(second green) = $\frac{4}{51}$ + $\frac{16}{153}$ + $\frac{2}{51}$ = $\frac{2}{9}$

73)

C. Find the hypotenuse of ΔQPT.

$H_1 = QT = \sqrt{3^2 + 2^2} = 3.6$

Find the diagonal of $QRST$.

$H_1 = SQ = \sqrt{3.6^2 + 3.6^2} = \mathbf{5.09}$

74)

B. Calculate the volume of water in tank A.

$V = l \times w \times h$

$5 \times 10 \times 1 = 50 \text{ ft}^3$

Find the height this volume would reach in tank B.

$V = l \times w \times h$

$50 = 5 \times 5 \times h$

$h = \mathbf{2\ ft}$

75)

D. Split the absolute value inequality into two inequalities and simplify. Switch the inequality when making one side negative.

$\frac{x}{8} \geq 1$

$x \geq 8$

$-\frac{x}{8} \geq 1$

$\frac{x}{8} \leq -1$

$x \leq -8$

$x \leq -8$ or $x \geq 8$ → $\mathbf{(-\infty, -8] \cup [8, \infty)}$

76)

B. The slope 0.0293 gives the increase in passenger car miles (in billions) for each year that passes. Muliply this value by 5 to find the increase that occurs over 5 years: $5(0.0293) =$ **0.1465 billion miles**.

77)

C. The greatest distance will be between two points at opposite ends of each sphere's diameters. Find the diameter of each sphere and add them.

$36\pi = \frac{4}{3}\pi r_1^3$

$r_1 = 3$

$d_1 = 2(3) = 6$

$288\pi = \frac{4}{3}\pi r_2^3$

$r_2 = 6$

$d_2 = 2(6) = 12$

$d_1 + d_1 = 6 + 12 = \mathbf{18}$

78)

A. The ellipse is horizontal, so the vertices will be the points $(h \pm a, k) =$ **(8 ± 10, 4)**.

79)

A. XW and XY are both radii, so $XW = XY = WY$.

Because ΔWXY is equilateral, $m\angle WXY = 60°$

The tangent line and the radius of the circle create a 90° angle, so $m\angle WXY = 90°$.

$\triangle ZXW$ is a 30–60–90 triangle, and $\angle WZX = \textbf{30}°$.

80)

B. The altitude creates a **30–60–90 triangle with a height of $5\sqrt{3}$.**

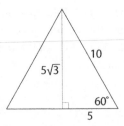

81)

A. Use the exponential decay function. The value $\frac{y}{a}$ represents the percent of the water remaining.

$y = a(1 - r)^t$

$y = (1 - 0.04)^7$

$y = (0.96)^7 = 0.751$

After 7 days, the new amount of water is 0.751, or 75.1% of the original amount. It has decreased by $100 - 75.1 = \textbf{24.9\%}$.

82)

C. The player earns money on either an 8 or one of the other even numbers.

A $2 bet has a –$2 value if the player gets an odd number, which has a probability of $\frac{5}{10}$.

A $2 bet has a $8 value if the player gets 8, which has a probability of $\frac{1}{10}$.

A $2 bet has a value of $1 for the other even numbers, which has a probability of $\frac{4}{10}$.

$-2\left(\frac{5}{10}\right) + 8\left(\frac{1}{10}\right) + 1\left(\frac{4}{10}\right) = \frac{2}{10} = $ **$0.20**

83)

B. Use the formula for the determinant of a 2 × 2 matrix.

$|\mathbf{D}| = ad - bc$

$= 3(9) - 7(4) = \textbf{-1}$

84)

C. Use the equation for tangent.

$\tan 50° = \frac{x}{300}$

$x = 300(\tan 50°)$

$x \approx \textbf{357}$

85)

A. Incorrect. Such a video would be a good alternative to a summative unit assessment for this project, but it is not the best answer choice.

B. Incorrect. A portfolio such as this would be a good alternative to a summative unit assessment for this project, but there is a better answer choice here.

C. Incorrect. An oral presentation summarizing the project and the ultimate learning outcomes would be a good alternative to a summative unit assessment, but this is not the best answer choice.

D. **Correct.** All of these alternative assessment strategies would be appropriate for a project-based unit.

86)

C. Write a system of equations and solve using substitution.

$5a + b - 2 = 180$

$b = \frac{7}{8}a$

$5a + \frac{7}{8}a - 2 = 180$

$a = \textbf{30.98}$

87)

D. Use the equation to find the nth term of an arithmetic sequence.

$a_1 = -57$

$d = -40 - (-57) = 17$

$n = 12$

$a_n = a_1 + d(n - 1)$

$a_{12} = -57 + 17(12 - 1)$

$a_{12} = \textbf{130}$

88)

D. For the function $y = a|x - h| + k$:

When $|a| > 1$, the graph will narrow.

When a is negative, the graph is reflected over the x-axis.

89)

A. Incorrect. If students are learning about area and perimeter for the first time, they will be uncertain of what the most important thing is.

B. **Correct.** Asking students to draw a model will help them analyze the definitions and visualize the meaning of area and perimeter.

C. Incorrect. Asking students to come up with a general formula should be covered after they have developed a better understanding.

D. Incorrect. Again, students must develop a better understanding before they are asked to come up with a general formula.

90)

D. Complete the square to put the quadratic equation in vertex form.

$y = 2x^2 + 12x - 3$

$y = 2(x^2 + 6x + \underline{\quad}) - 3 + \underline{\quad}$

$y = 2(x^2 + 6x + 9) - 3 - 18$

$y = 2(x + 3)2 - 21$

91)

C. Use the formula for the inverse of a 2 × 2 matrix.

$\begin{bmatrix} a & b \\ c & d \end{bmatrix}^{-1} = \dfrac{1}{ad-bc}\begin{bmatrix} d & -b \\ -c & a \end{bmatrix}$

$= \dfrac{1}{-1}\begin{bmatrix} 9 & -7 \\ -4 & 3 \end{bmatrix}$

$= \begin{bmatrix} -9 & 7 \\ 4 & -3 \end{bmatrix}$

92)

A. Incorrect. This question requires an overly specific answer; it is not a guiding question.

B. Incorrect. Again, this question is too specific.

C. **Correct.** Asking this question will remind students how a linear equation can be graphed and help form a connection between graphing lines and graphing parabolas.

D. Incorrect. While this question could help the student visualize the parabola, it is not the best answer choice.

93)

A. Set the numerator of the resulting rational function equal to 0 to find the zeros.

$\left(\dfrac{g}{h}\right)(k) = \dfrac{(-3k^2 - k)}{(-2k - 4)}$

$0 = 3k^2 - k$

$0 = k(3k - 1)$

$k = 0, \dfrac{1}{3}$

94)

A. A rotation of 180° is found by performing the transformation $(x, y) \to (-x, -y)$. Since this rotation is around the point A, treat point A as the origin.

A remains **(1, 1)**.

B is 4 units right and 2 units up from A, so count 4 units left and 2 units down from A to find B': (5, 3) → **(−3, −1)**

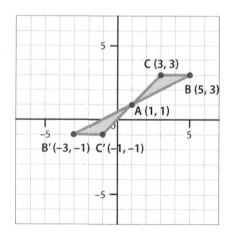

C is 2 units right and 2 units up from A, so count 2 units left and 2 units down from A to find C': $(3, 3)$ →
(−1, −1)

95)

C.   Find the height of the cylinder using the equation for surface area.

$SA = 2\pi rh + 2\pi r^2$

$48\pi = 2\pi(4)h + 2\pi(4)^2$

$h = 2$

Find the volume using the volume equation.

$V = \pi r^2 h$

$V = \pi(4)^2(2) = \mathbf{32\pi \ ft.^3}$

96)

D.   Find the amount of change and add to the original amount.

amount of change = original amount × percent change

$= 37{,}500 \times 0.055 = 2{,}062.50$

$37{,}500 + 2{,}062.50 = \mathbf{\$39{,}562.50}$

97)

A.   In 2012, New York had more months with less than 3 inches of rain than either Chicago or Houston.

98)

A.   I. True: The row "25 − 35 year-olds" and the column "more than 4 hours" show a relative frequency of 0.20 or 20%.

II. True: The total relative frequency of participants who spend 2 to 4 hours on social media a day is 0.26 or 26%.

III. False: These percentages cannot be added because the "whole" is not the same. To calculate the percentage of people aged 15 − 35 who use social media less than 2 hours a day, find the total number of people in each category and divide by the total number of people in both categories.

people 15 − 25 years old spending < 2 hrs. $= 0.15(200) = 30$

people 25 − 35 years old spending < 2 hrs. $= 0.52(200) = 104$

percentage of people 15 − 35 years old spending < 2 hrs. $= \frac{30 + 104}{400} = 0.34 = $ 34%

IV. False: The number of people aged 35 − 45 who used social media more than 4 hours a day is $0.05(200) = 10$.

99)

C.   Use the equation for exponential decay to find the year the population reached 1 million.

$y = a(1 - r)^t$

$y = 1250000(1 - 0.05)^t$

$1{,}000{,}000 = 1250000(1 - 0.05)^t$

$0.8 = 0.95^t$

$\log_{0.8} 0.8 = \log_{0.8} 0.95^t$

$1 = t \times \frac{\log 0.95}{\log .8}$

$t = 4.35$

The population will reach 1 million in 2013.

100)

A.   Rewrite the equation in exponential form and solve.

$\log_7(-2z) = 0$

$7^0 = -2z$

$1 = -2z$

$z = -\frac{1}{2}$

Go to **www.cirrustestprep.com/texes-math-online-resources** to access your second TExES Mathematics practice test and other online study resources.

Made in the USA
Coppell, TX
22 October 2023